AF800554

Lecture Notes in Mechanical Engineering

Series Editors

Fakher Chaari, *National School of Engineers, University of Sfax, Sfax, Tunisia*
Francesco Gherardini, *Dipartimento di Ingegneria "Enzo Ferrari", Università di Modena e Reggio Emilia, Modena, Italy*
Vitalii Ivanov, *Department of Manufacturing Engineering, Machines and Tools, Sumy State University, Sumy, Ukraine*
Mohamed Haddar, *National School of Engineers of Sfax (ENIS), Sfax, Tunisia*

Editorial Board Members

Francisco Cavas-Martínez, *Departamento de Estructuras, Construcción y Expresión Gráfica Universidad Politécnica de Cartagena, Cartagena, Spain*
Francesca di Mare, *Institute of Energy Technology, Ruhr-Universität Bochum, Bochum, Germany*
Young W. Kwon, *Department of Manufacturing Engineering and Aerospace Engineering, Graduate School of Engineering and Applied Science, Monterey, USA*
Tullio A. M. Tolio, *Department of Mechanical Engineering, Politecnico di Milano, Milano, Italy*
Justyna Trojanowska, *Poznan University of Technology, Poznan, Poland*
Robert Schmitt, *RWTH Aachen University, Aachen, Germany*
Jinyang Xu, *School of Mechanical Engineering, Shanghai Jiao Tong University, Shanghai, China*

Lecture Notes in Mechanical Engineering (LNME) publishes the latest developments in Mechanical Engineering—quickly, informally and with high quality. Original research or contributions reported in proceedings and post-proceedings represents the core of LNME. Volumes published in LNME embrace all aspects, subfields and new challenges of mechanical engineering.

To submit a proposal or request further information, please contact the Springer Editor of your location:

Europe, USA, Africa: Leontina Di Cecco at Leontina.dicecco@springer.com
China: Ella Zhang at ella.zhang@cn.springernature.com
India, Rest of Asia, Australia, New Zealand: Swati Meherishi at swati.meherishi@springer.com

Topics in the series include:

- Engineering Design
- Machinery and Machine Elements
- Mechanical Structures and Stress Analysis
- Automotive Engineering
- Engine Technology
- Aerospace Technology and Astronautics
- Nanotechnology and Microengineering
- Control, Robotics, Mechatronics
- MEMS
- Theoretical and Applied Mechanics
- Dynamical Systems, Control
- Fluid Mechanics
- Engineering Thermodynamics, Heat and Mass Transfer
- Manufacturing Engineering and Smart Manufacturing
- Precision Engineering, Instrumentation, Measurement
- Materials Engineering
- Tribology and Surface Technology

Indexed by SCOPUS, EI Compendex, and INSPEC.

All books published in the series are evaluated by Web of Science for the Conference Proceedings Citation Index (CPCI).

To submit a proposal for a monograph, please check our Springer Tracts in Mechanical Engineering at https://link.springer.com/bookseries/11693.

John Mo · Parames Chutima
Editors

Proceedings of The 7th International Conference on Mechanical, System and Control Engineering

Proceedings of ICMSC 2024
20–22 November, 2024

Springer

Editors
John Mo
School of Engineering
RMIT University
Melbourne, VIC, Australia

Parames Chutima
Department of Industrial Engineering,
Faculty of Engineering
Chulalongkorn University
Pathum Wan, Thailand

ISSN 2195-4356 ISSN 2195-4364 (electronic)
Lecture Notes in Mechanical Engineering
ISBN 978-981-96-5379-9 ISBN 978-981-96-5380-5 (eBook)
https://doi.org/10.1007/978-981-96-5380-5

© The Editor(s) (if applicable) and The Author(s), under exclusive license
to Springer Nature Singapore Pte Ltd. 2025

This work is subject to copyright. All rights are solely and exclusively licensed by the Publisher, whether the whole or part of the material is concerned, specifically the rights of translation, reprinting, reuse of illustrations, recitation, broadcasting, reproduction on microfilms or in any other physical way, and transmission or information storage and retrieval, electronic adaptation, computer software, or by similar or dissimilar methodology now known or hereafter developed.
The use of general descriptive names, registered names, trademarks, service marks, etc. in this publication does not imply, even in the absence of a specific statement, that such names are exempt from the relevant protective laws and regulations and therefore free for general use.
The publisher, the authors and the editors are safe to assume that the advice and information in this book are believed to be true and accurate at the date of publication. Neither the publisher nor the authors or the editors give a warranty, expressed or implied, with respect to the material contained herein or for any errors or omissions that may have been made. The publisher remains neutral with regard to jurisdictional claims in published maps and institutional affiliations.

This Springer imprint is published by the registered company Springer Nature Singapore Pte Ltd.
The registered company address is: 152 Beach Road, #21-01/04 Gateway East, Singapore 189721, Singapore

If disposing of this product, please recycle the paper.

Preface

This book contains the final version of papers accepted and presented in the 7th International Conference on Mechanical, System and Control Engineering (ICMSC 2024) which was held in the stylish and spectacular city of Bangkok, Thailand during November 20–22, 2024. The conference brought together some of the latest research and scientific results of academia scholars and practising engineers working in the fields related to Mechanical, Systems and Control topics. This is a face-to-face conference hosted by one of the reputable universities in Bangkok, the Chulalongkorn University. Like previous years, all papers submitted to the conference have been thoroughly peer reviewed by conference committee members and international experts.

The long history of ISMSC has proven that it is importance in the promotion of research outcomes in the conference fields. Delegates from China and Taiwan, Japan, Thailand, Saudi Arabia, Czech, Russia, Kazakhstan and Korea attended two days of activities. Apart from enjoying friendly campus environment of Chulalongkorn University and excellent local foods, conference participators had lots of time to exchange research ideas, outcomes and aspirations.

The manuscripts collected in this ICMSC 2024 Proceedings are final versions after authors responded and revised according to reviewer comments. The conference proceedings can be divided into four chapters: (1) Functional Material Design, (2) Mechanical Structure Design and Analysis, (3) Micro Mechanical System Design and (4) Intelligent Unmanned System.

Chapter 1 has 3 papers related to properties of engineering materials in application environments. Paper by Wang et al reported a prolong experiment testing corrosion behaviour of galvanised steel in marine environment. Their results indicated that corrosion occurred at contact points. This paper is a valuable reference for managing corrosion in harsh environment such as marine applications. Paper by Kumar et al reported another valuable experiment optimising composition and formation process of copper matrix composites. Copper matrix composites are increasingly used in aerospace, automotive and other engineering systems. The results from this paper can be applied directly to improve strength to weight performance of new products. Longevity of steel in geothermal production facilities was examined by Wolf and Pfenning. Corrosion effects in geothermal systems are more severe than marine environments. This paper reported a novel signal-based technique of predicting material strength changes. Their findings provide valuable reference to new methods of early failure prediction such as crack in the geothermal system. Papers in Chapter 1 contribute valuable new knowledge of understanding and processing of materials, paving the way to design more robust and reliable mechanical systems.

Chapter 2 has 4 papers relating to mechanical structure design. Paper by Guo et al proposed a novel virtual spring constrained method to analysis flexible machines. Their new method represents a break-through to solving non-convergence issues in numerical analysis of connected mechanical structures and is able to eliminate uncertainty in mechanical system simulation. Paper by Gong et al reported an innovative design that could correct mini-satellites orbits. Due to limited resources, mini-satellites can't use traditional methods of de-orbiting. The new design made use of changing mechanical characteristics of the satellite to adjust its flight path and has been proven by simulation. Another interesting paper by Liu et al presents design of an unlocking device that minimised impact during separation of satellites from carrying system. Significant improvements in terms of power consumption, response time and loading were verified with numerical analysis and targeted experiments. Paper by Toyoshima et al focused on the design of mechanisms assisting rehabilitation of human lower body. The system used a single measurement unit with deep learning algorithms to monitor and predict lower body movements to enable more accurate real time assistance to people in need. Papers in this chapter demonstrate great potentials in cost savings and efficiency with cleverly designed mechanical structures.

Chapter 3 has 4 papers relating smart mechanical system design. Wild boar activities have caused substantial losses to farmers. A new animal recognition system has been created by Niimoto el al to identify boar presence early so that appropriate control measure can be developed in time. Their system applied a new artificial intelligence model and improved recognition accuracy and speed of analysis greatly. Paper by Thiemsakul et al investigated best practices of separating microplastics from freshwater and marine ecosystems. They proposed to use hydrocyclones and were successful in demonstrating effectiveness of the new system in separating different types of microplastics. Senin et al developed a neural network algorithm to analyse surface roughness in preparation for smooth milling. The system was preset machining parameters to ensure best manufacturing quality of the final product. Paper by Khan et al investigated the effect of steam jets impinging on walls. Steam jets have many applications in ventilation, petrochemical, compressors and so on. This paper provides an in-depth analysis of steam jets and subsequent fluid conditions, supporting more efficient design of steam jets. Papers in Chapter 3 demonstrate the broad range of research fields that can be applied to control of mechanical systems.

Chapter 4 has 3 papers focusing on operations of unmanned systems. Alshamlan et al discussed the importance of monitoring conditions on satellites. Rapid changing space conditions can affect performance of satellites. Their specialised attitude determination and control system was able to monitor temperatures on board a low-orbit satellite within one degree over the designed operating range. Uesugi et al developed a light-weight guard for indoor drone operations. The guard not only protected the drone, people in the room as well as assisting the drone to recover after collision or falls. The system could be used to expand indoor applications of drone operations. Working in parallel, Nagata et al developed a flight path generation algorithm for indoor drone operations. Drons can reach difficult to access locations inside buildings and structure in any environments. Availability of a clever flight path generation system significantly improve efficient of internal inspections of infrastructure in indoor environments. Papers in Chapter 4 are good examples of state-of-the-art development of enhancing capabilities of mechanical systems.

I would like to sincerely thank contributors, reviewers, organizing committee members of ICMSC 2024 for providing a great conference platform to engage discussions among the participants. ICMSC 2024 is the seventh in the conference series. With continuous support and collaboration of contributors, reviewers, committee members, the conference series will continue in the future and will again provide an effective platform for exchange of advanced know-how and knowledge, while also fostering potential international research collaborations in the topics of Mechanical, Systems and Control Engineering.

November 2024

John Mo
Conference Chair, ICMSC 2024

Organization

Conference Chairs

John Mo	Royal Melbourne Institute of Technology, Australia
Parames Chutima	Chulalongkorn University, Thailand

Conference Co-chair

Ratchatin Chancharoen	Chulalongkorn University, Thailand

Program Chairs

Dulyachot Cholaseuk	Thammasat University, Thailand
Chiang-Ju Chien	Huafan University, Taiwan
Theeraphong Wongratanaphaisan	Chiang Mai University, Thailand

Program Co-chairs

Bovornchok Poowat	King Mongkut's University of Technology Thonburi, Thailand
Thongchai Fongsamootr	Chiang Mai University, Thailand
Ben Niu	Dalian University of Technology, China

Publicity Committee Chairs

Bunyong Rungroungdouyboon	Thammasat University, Thailand
Monkol Seenawat	King Mongkut's University of Technology Thonburi, Thailand

Technical Committee

Chanawee Promaue	University of Bristol, UK
Nopparat Seemuang	KMUTNB, Thailand
Shyh-Chour Huang	National Kaohsiung University of Science and Technology, Taiwan
Thananchai Leephakpreeda	Thammasat University, Thailand
Gheorghe-Daniel Andreescu	Universitatea Politehnica Timisoara, Romania
Yuri N. Skiba	Universidad Nacional Autónoma de México, Mexico
Valery Matveev	Samara National Research University, Russia
Alexander Kusumov	Kazan National Technical University, Russia
Mohamed M. Hafez	University of California, USA
Mikhail Murashov	Bauman Moscow State Technical University, Russia
Motahar Reza	GITAM Deemed to be University, India
Alexander Kuzmin	St Petersburg State University, Russia
Vu Ngoc Pi	Nguyen Tat Thanh University, Vietnam
Khaled Abou-El-Hossein	Nelson Mandela University, South Africa
Yoshifumi Yokoi	National Defense Academy of Japan, Japan
Ruchkin Vladimir Nikolaevich	Ryazan State University named for S. Yesenin, Russia
Faris Tarlochan	Qatar University, Qatar
Yaguang Zhu	Chang'an University, China
Vadim Kramar	Sevastopol State University, Russia
Andrzej Gontarz	Lublin University of Technology, Poland
Liangen Yang	Hubei University of Technology, China
Alexey Zhdanov	Vladimir State University, Russia
Eugene Larkin	Tula State University, Russia
Valeriy Matveev	Samara National Research University, Russia
Jürgen Weber	Institute of Fluid Power, Germany
Sumit Das Lala	Parul University, India
Payel Deb	Parul University, India
Anand Joshi	Parul University, India
Omprakash Bharti	Parul University, India
Prince Jain	Parul University, India

Contents

Functional Material Design

Study on Corrosion Behavior of 300M Ultra-high Strength Steel-5228A/CCF300 Composite Galvanic Connectors in Marine Atmospheric Environment

Jin-mei Wang[1], Lai-zheng Luo[1,2](✉), Wei Zhang[1], Li-jun Feng[1], and Jian-hong Liu[1]

[1] Southwest Technology and Engineering Research Institute, Chongqing 400039, China
280924035@qq.com

[2] College of Chemistry and Chemical Engineering, Chongqing University, Chongqing 400030, China

Abstract. Exposure tests were carried out on 300M steel cadmium plated titanium, 300M steel cadmium plated titanium-5228A/CCF300 composite galvanic connectors and 300M steel coating- 5228A/CCF300 composite galvanic connectors at Wanning test station to study the corrosion behavior of galvanic couple connectors composed of 300M steel and 5228A/CCF300 composite. The results showed that after two years of exposure to the marine atmospheric environment, significant galvanic corrosion occurred at the contact area of 300M steel cadmium plated titanium- 5228A/CCF300 composite galvanic connectors. The tensile strength and elongation after fracture decreased by 6.8% and 39.5% due to galvanic corrosion damage. No obvious corrosion was found in 300M steel cadmium plated titanium and 300M steel coating-5228A/CCF300 composite galvanic connectors, and the tensile properties did not show a significant decrease. Hydrogen evolution reaction will occur on the surface of the composite material, and the anode dissolution of Fe and reduction of O will occur in the structural steel when the structural steel and composite material connection structures are used in the marine atmosphere. In order to avoid the occurrence of galvanic corrosion, various protective measures such as surface treatment, organic coating and sealant should be adopted for low-potential structural steel.

Keywords: 300M ultra-high strength steel · 5228A/CCF300 composite · Marine atmospheric environment · Corrosion behavior

1 Introduction

Carbon fiber composite is a composite material with resin as matrix and carbon fiber as reinforcement [1]. Compared with traditional metal materials such as aluminum alloy and alloy steel, carbon fiber composites have many advantages such as high specific strength, high specific modulus, corrosion resistance and fatigue resistance. When obtaining the same strength, function and other comprehensive properties as the metal structure, its

© The Author(s), under exclusive license to Springer Nature Singapore Pte Ltd. 2025
J. Mo and P. Chutima (Eds.): ICMSC 2024, LNME, pp. 3–12, 2025.
https://doi.org/10.1007/978-981-96-5380-5_1

weight can be reduced by 25%–50% [2]. Since the carbon fiber composite material is widely used in military and civil fields, especially in the aerospace field, its excellent characteristics make it become an indispensable and rare multi-functional special engineering material. Nowadays, carbon fiber composite materials have become the four major structural materials in the field of aviation and aerospace together with aluminum alloy, titanium alloy and alloy steel [3], which has become a symbol to measure the upgrading of products in a national security strategy and related fields of national economy.

Carbon fiber composites have good electrical conductivity and positive electrode potential because they retain the electrochemical properties of carbon itself. After carbon fiber composites are connected to metal materials in corrosive media, the corrosion rate of metal materials with negative electrode potential will be accelerated, that is, galvanic corrosion occurs when carbon fiber composite materials are in contact with metal materials [4]. Lu et al. [5] studied the galvanic corrosion and control methods of composite materials and metals in atmospheric environment. The results showed that in the atmospheric environment of Beijing, Qingdao Tuandao and Hainan Wanning, when carbon fiber composite materials were coupled with 30CrMnSiA steel, LF2 anti-rust aluminum and TC4 titanium alloy, the climatic environment of Hainan Wanning had the most significant effect on the galvanic corrosion of composite materials and 30CrMnSiA steel.

300M ultra-high strength steel is a new type of low alloy ultra-high strength steel developed by American International Nickel Company in 1952, which is based on 4340 steel and added about 1.5% Si [6]. 300M steel is a kind of medium-carbon low-alloy ultra-high strength steel widely used in the aviation field at home and abroad. It has good toughness and fatigue properties. Different microstructures can be obtained by different heat treatments to achieve different strength levels. It is mainly used to manufacture aircraft landing gears and important structural parts. During the use of 300M steel, it is inevitable to contact with dissimilar metal materials or composite materials, which is prone to galvanic corrosion, especially in the marine atmospheric environment. At present, there are few reports on the corrosion behavior of the contact between 300M steel and carbon fiber composite materials. In order to ensure the reliable use of the long-term contact between 300M steel and carbon fiber composite materials in the marine environment, the atmospheric exposure tests on 300M steel and 5228A/CCF300 carbon fiber composite materials galvanic connectors were carried out at Hainan Wanning in this paper. The corrosion characteristics and mechanical properties degradation of composite and structural steel galvanic connectors were studied, and the causes of galvanic corrosion when the two were in contact were preliminarily discussed, which provided important reference value for the engineering application of carbon fiber composite and ultra-high strength steel connection structure.

2 Test

2.1 Test Samples

The test samples include three kinds, which are 300M steel cadmium plated titanium plate tensile sample, 300M steel cadmium plated titanium-5228A/CCF300 composite galvanic connectors and 300M steel coating-5228A/CCF300 composite galvanic connectors. The 300M steel in the galvanic connector is the anode material, and the 5228A/CCF300 composite material is the cathode material. The anode material and the cathode material are isolated by XM-33 two-component polysulfide sealant. The basic composition of the galvanic connector is shown in Table 1, and the physical photo is shown in Fig. 1.

Table 1. Basic composition of three test samples.

Test samples	Anode material	Cathode material	Isolation measures
300M steel cadmium plated titanium-5228A/CCF300 composite galvanic connectors	300M steel + cadmium plated titanium	5228A/CCF300 composite	XM-33 two-Component polysulfide sealant
300M steel coating-5228A/CCF300 composite galvanic connectors	300M steel + cadmium plated titanium + H06-27 primer + 13-2 topcoat	5228A/CCF300 composite	XM-33 two-Component polysulfide sealantt

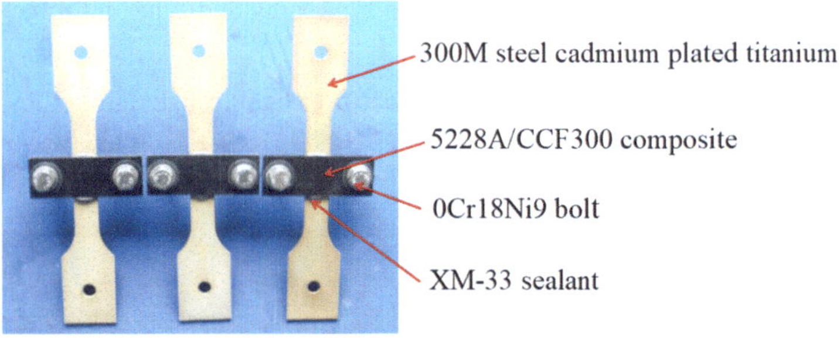

Fig. 1. The physical diagram of the galvanic connector

2.2 Marine Atmospheric Environment Test

Test Environment. The test environment is the Hainan Wanning test station with hot and humid marine atmospheric environment. It has the characteristics of high temperature, high humidity, high salt fog and strong solar radiation. It is a serious area of marine atmospheric environment corrosion.

Exposure Test. The marine atmospheric environment exposure test was carried out according to GB/T 14165–2008 ≪Corrosion of metals and alloys—Atmospheric corrosion testing—General requirements for field tests≫. The samples were exposed southward and at an angle of 45° with the horizontal direction. The test period was 2 years. Three samples were taken from the test rack for tensile performance testing in the first and second years. The field test photos of the galvanic connector are shown in Fig. 2.

Fig. 2. Marine atmospheric environment exposure test.

2.3 Sample Performance Testing and Analysis

Macroscopic Corrosion Morphology. The last cycle sample was used to track and detect the macroscopic corrosion morphology. During the test, it was detected once every 3 months. The corrosion behavior of the contact part of the galvanic connector was observed, and the macroscopic corrosion type, area and other information were recorded and photographed.

Tensile Properties. According to GB/T 228–2010 ≪Metallic materials—Tensile test—Part 1: Method of test at room temperature≫, MTS880 electro-hydraulic servo material testing machine was used to carry out tensile test on the test samples at a tensile speed of 1.5 mm·min^{-1}.

3 Results and Discussion

3.1 Macroscopic Corrosion Characteristics

Corrosion Characteristics of 300M Steel Cadmium Plated Titanium. The macroscopic corrosion morphology of 300M steel cadmium plated titanium in marine atmospheric environment test for 2 years is shown in Fig. 3. It can be seen from the figure that with the extension of exposure time, the surface color of 300M steel cadmium plated titanium gradually becomes lighter, from yellowish brown to white, but the substrate has no obvious corrosion, indicating that the surface treatment layer of cadmium plated titanium still has a good protective function for 300M steel.

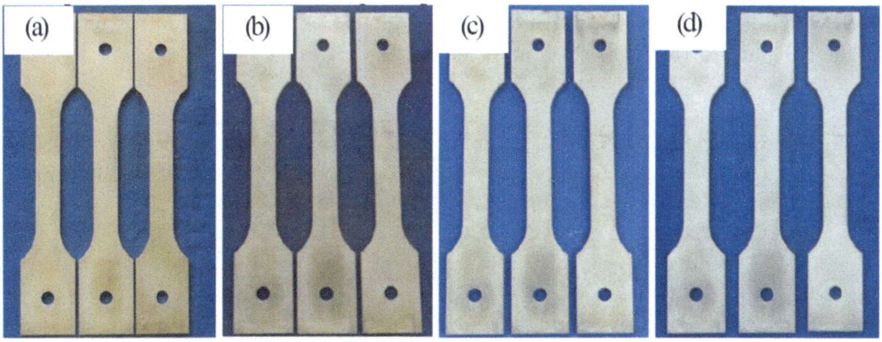

Fig. 3. Macroscopic corrosion morphology of 300M steel cadmium plated titanium: (a) Original; (b) 3 months; (c) 12 months; (d) 24 months.

Corrosion Characteristics of 300M Steel Cadmium Plated Titanium-5228A/CCF300 Composite Galvanic Connectors. The macroscopic corrosion morphology of the galvanic connector composed of 300M steel cadmium plated titanium and 5228A/CCF300 composite after 2 years of marine atmospheric environment test is shown in Fig. 4. It can be seen from Fig. 4(a) and Fig. 4(c), that the contact part of the galvanic connector was first corroded after 3 months of exposure to the marine atmospheric environment. It shows that when 300M steel contacts with composite materials, it is difficult to isolate the occurrence of galvanic corrosion only by cadmium-plated titanium treatment. With the extension of exposure time, the corrosion gradually developed from near to far, the color of corrosion products gradually deepened, and the degree and area of corrosion gradually increased. After 24 months of exposure, the color of corrosion products became dark brown, and the maximum length of corrosion at the contact site was about 35 mm. This is mainly due to the uneven distribution of the corrosion current in the anode area. The closer to the contact, the greater the corrosion current is. The farther away from the contact, the smaller the corrosion current is. Beyond a certain range, the galvanic corrosion effect is almost zero. This range is called ' effective distance ', which is related to the electromotive force of the corrosion cell, the conductivity of the solution, and the geometry of the contact.

It can be seen from Fig. 4(b) and Fig. 4(d) that there is a significant difference in the degree of corrosion between the upper contact part and the lower contact part. After exposure to marine atmospheric environment for 3 months, white and reddish-brown corrosion products appeared in the upper contact area, and white corrosion products appeared in the lower contact area. The white corrosion products indicated that the surface treatment layer of cadmium-plated titanium was corroded, but the corrosion did not penetrate into the 300M steel substrate. The reddish-brown corrosion products indicate that the 300M steel substrate has been corroded. With the extension of exposure time, the white corrosion products were all transformed into reddish-brown corrosion products, and eventually developed into dark brown. It can be seen that the corrosion of the upper contact part is obviously more serious than that of the lower contact part during the exposure of the marine atmospheric environment. This is mainly due to the 45° angle exposure of the galvanic connectors, and the upper contact part is easy to store water, which is more likely to form the ion conductive branch of the corrosion battery, and the galvanic corrosion effect time is longer.

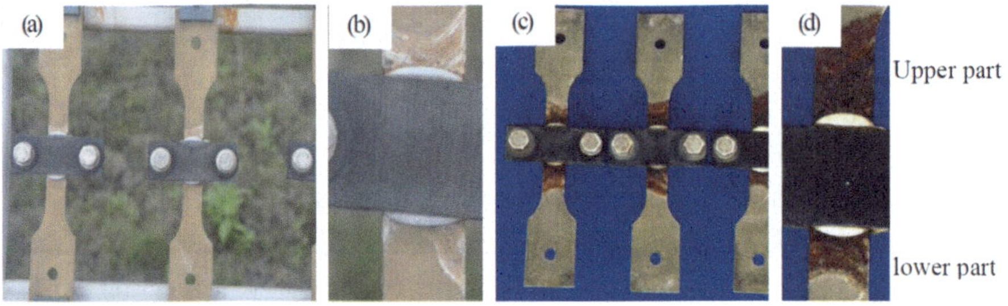

Fig. 4. Macroscopic corrosion morphology of 300M steel cadmium plated titanium-5228A/CCF300 composite galvanic connectors: (a) 3 months (full view); (b) 3 months (partial); (c) 24 months (full view); (d) 24 months (partial).

Corrosion Characteristics of 300M Steel Coating-5228A/CCF300 Composite Galvanic Connectors. The macroscopic corrosion morphology of the galvanic connector composed of 300M steel coating and 5228A/CCF300 composite after 2 years of marine atmospheric environment test is shown in Fig. 5. It can be seen from the figure that there is no obvious corrosion phenomenon in the contact part of the 300M steel coating-5228A/CCF300 composite galvanic connector during the exposure of the marine atmospheric environment. This indicates that 300M steel treated with cadmium plated titanium and organic coating can effectively isolate galvanic corrosion when in contact with 5228A/CCF300 composite materials.

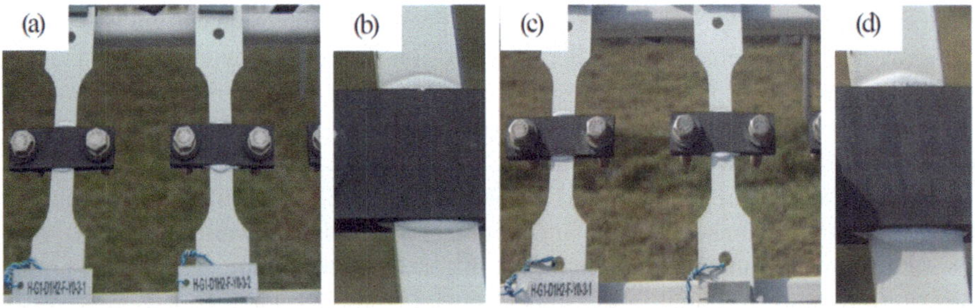

Fig. 5. Macroscopic corrosion characteristics of 300M steel coating-5228A / CCF300 composite galvanic connector: (a) 3 months (full view); (b) 3 months (partial); (c) 24 months (full view); (d) 24 months (partial).

3.2 Tensile Properties

The tensile properties curves of 300M steel cadmium plated titanium, 300M steel cadmium plated titanium-5228A/CCF300 composite galvanic connector and 300M steel coating-5228A/CCF300 composite galvanic connector are shown in Fig. 6. It can be seen from the Fig. 6(a), that the tensile strength of 300M steel cadmium plated titanium-5228A/CCF300 composite galvanic connector decreases significantly with the prolongation of exposure time in the marine atmospheric environment. After 2 years of exposure, the tensile strength of the 300M steel cadmium plated titanium-5228A/CCF300 composite galvanic connector decreased from 1835 MPa to 1710 MPa, and the tensile strength decreased by 6.8%. During the exposure of 300M steel coating-5228A/CCF300 composite galvanic connector and 300M steel cadmium plated titanium in marine atmospheric environment, the tensile strength did not decrease significantly with the prolongation of exposure time. After 2 years of exposure, the tensile strength decreased from 1835 MPa to 1811 MPa and 1829 MPa, and the tensile strength decreased by only 1.3% and 0.3% respectively. The results show that when 300M steel contacts with 5228A/CCF300 composite, although the surface of 300 steel is treated with cadmium plated titanium and the contact part is coated with sealant, the tensile strength decreases obviously. The results of tensile strength of 300M steel coating-5228A/CCF300 composite galvanic connectors show that 300M steel can effectively prevent the occurrence of galvanic corrosion and greatly reduce the decrease of tensile strength after cadmium plated titanium and protective coating treatment.

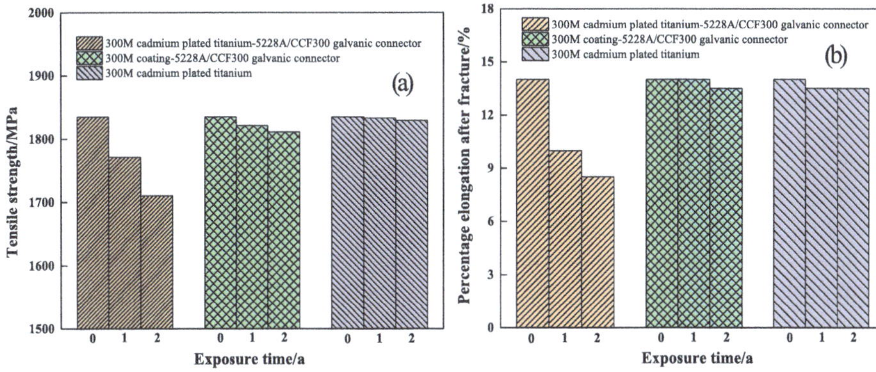

Fig. 6. Comparison of tensile properties of three samples: (a) tensile strength; (b) percentage elongation after fracture.

It can be seen from the Fig. 6(b), that the percentage elongation after fracture of 300M steel cadmium plated titanium-5228A/CCF300 composite galvanic connector decreases significantly with the prolongation of exposure time in the marine atmospheric environment. After 2 years of exposure, the percentage elongation after fracture of 300M steel cadmium plated titanium-5228A/CCF300 composite galvanic connector decreased from 14% to 8.5%, and the decrease rate of percentage elongation after fracture was 39.5%. During the exposure of 300M steel coating-5228A/CCF300 composite galvanic connector and 300M steel cadmium plated titanium in marine atmospheric environment, the percentage elongation after fracture did not decrease significantly with the prolongation of exposure time. After 2 years of exposure, the percentage elongation after fracture decreased from 14% to 13.5%, and the decrease rate of percentage elongation after fracture only decreased by 3.5%. It can be seen that the variation trends of tensile strength and percentage elongation after fracture of the three samples are basically the same, and the results of tensile properties further confirm the results of macroscopic corrosion characteristics.

3.3 Causes and Mechanism of Galvanic Corrosion

The occurrence of galvanic corrosion must have the following three basic conditions. The first is the presence of corrosive media. Wanning test station in Hainan has the characteristics of hot and humid marine atmospheric environment such as high temperature, high humidity and high salt fog. Among them, the annual average relative humidity is as high as 86%, the Cl^- content in the atmospheric environment is as high as 0.48 mg/(100 cm^2·d), the pH value of precipitation is 5.4 [7], and there is an ion conduction branch required for galvanic corrosion. Secondly, there are two or more different metals or non-metals with potential differences. The 5228A/CCF300 carbon fiber composite has a higher potential, while the 300M ultra-high strength steel has a lower potential, and there is a large potential difference between the two contacts. The third is conductive connection. Although the contact surface between 300M cadmium plated titanium and 5228A/CCF300 composite is isolated by sealant, because the relative humidity of the environment is as high as 86%, a continuous thin liquid film will be formed on the surface of the sealant due to the high relative humidity. It connects 300M steel cadmium plated titanium with 5228A/CCF300 composite to form an electronic conductive branch of corrosion battery [8], which creates conditions for galvanic corrosion.

In the 300M steel cadmium plated titanium-5228A/CCF300 composite galvanic connector, the 300M steel cadmium plated titanium is the anode material, and the 5228A/CCF300 composite is the cathode material. The corrosion potential of the cadmium plated titanium layer (−749.28 mV) is significantly lower than that of the bare 300M steel (−639.48 mV), which belongs to the anodic coating. In the marine atmospheric environment, the coating preferentially corrodes to form Cd $(OH)_2$ and CdO_2, which plays a role in protecting the matrix. When the corrosive medium in the environment penetrates into the pores of the coating, the corrosion of the anode coating is caused by the formation of a corrosion galvanic cell with a large anode and a small cathode between the coating and the substrate. As the corrosion progresses, the cathode area gradually increases, and the 300M steel substrate is gradually exposed and corroded, which will cause the anodic dissolution of Fe and the reduction of O [9]:

$$Fe - 2e^- \rightarrow Fe^{2+} \tag{1}$$

$$O_2 + 2H_2O + 4e^- \rightarrow 4OH^- \tag{2}$$

Under the condition of sufficient O_2, Fe^{2+} further reacts:

$$Fe^{2+} + 2OH^- \rightarrow Fe(OH)_2 \tag{3}$$

$$3Fe(OH)_2 + 1/2O_2 \rightarrow Fe_3O_4 + 3H_2O \tag{4}$$

As a cathode material, 5228A/CCF300 carbon fiber composite is an electronic conductive phase and Cl^- is an ionic conductive phase. In acidic media, hydrogen evolution reaction mainly occurs on the surface of 5228A/CCF300 carbon fiber composite:

$$2H^+ + 2e^- \rightarrow H_2 \uparrow \tag{5}$$

$$O_2 + 4H^+ + 4e^- \rightarrow 2H_2O \tag{6}$$

During the exposure of marine atmospheric environment, the electrode reaction process on the surface of 5228A/CCF300 carbon fiber is very complicated, and these redox reactions are controlled by O_2 concentration difference. The reaction rate is controlled by the process of dissolved O_2 diffusing to the surface of carbon fiber [10].

4 Conclusion

(1) After being exposed to the hot and humid marine atmospheric environment for 2 years, the surface color of 300M steel cadmium plated titanium only became lighter, and no significant corrosion phenomenon was observed. The tensile properties also did not show a significant decrease, indicating that the surface treatment layer of cadmium plated titanium has a good protective function for 300M steel.

(2) After exposure to the hot and humid marine atmosphere for 3 months, the reddish-brown corrosion products began to appear at the contact part of the 300M steel cadmium plated titanium-5228A/CCF300 composite galvanic connector. With the prolongation of exposure time, the galvanic corrosion phenomenon gradually increased. After 2 years of exposure, galvanic corrosion damage led to a significant decrease in tensile properties, and the tensile strength and elongation decreased by 6.8% and 39.5%, respectively.

(3) After 2 years of exposure to hot and humid marine atmospheric environment, there was no obvious corrosion in the contact part of the 300M steel coating-5228A/CCF300 composite galvanic connector, and the tensile properties did not decrease significantly, indicating that 300M steel can effectively isolate the occurrence of galvanic corrosion after being treated with cadmium plated titanium and sprayed with organic coating.

(4) Hydrogen evolution reaction will occur on the surface of the composite material, and the anodic dissolution of Fe and the reduction of O will occur on the structural steel when galvanic connector is exposed to the hot and humid marine atmospheric environment. Therefore, in order to avoid galvanic corrosion, the contact surface of the 300M steel coating and 5228A/CCF300 composite is isolated by insulating sealant, the structural steel should also be comprehensively protected by surface treatment and spraying organic coatings.

References

1. Zhang, J., Sheng, Z., Sun, J.: Design Handbook of Composites Constructure. Beijing Yuhang Publishing Company, Beijing (1999)
2. Li, W., Guo, Q.: Application of carbon fiber composites to cosmonautic fields. Chin. J. Opt. **4**(3), 201–212 (2011)
3. Chen, S.: Design Handbook of Composite Material. Aviation Industry Press, Beijing (1990)
4. Tuker, W.C., Brown, R., Russell, L.: Corrosion between a graphite/polymer composite and metals. Compos. Mater. **1**, 92–102 (1990)
5. Lu, F., Zhong, Q., Cao, C.: Galvanic corrosion and controlling of GECM and metals in atmospheric environment. Mater. Prot. **35**(12), 19–22 (2002)
6. Long, H., Min, H.: Fatigue property of 300M steel under marine atmospheric. J. Xi'an Technol. Univ. **34**(10), 824–829 (2014)
7. Luo, L., Zhou, K., Li, X., et al.: Corrosion damage behavior of 30CrMnSiNi2A steel under coupling action of marine atmospheric environment and tensile fatigue load. Surf. Tech. **58**, 349–358 (2021)
8. Masatoshi, S., Yukiya, S.: Electrochemical noise analysis of galvanic corrosion of anodized aluminum in chloride environments. Electrochemistry **74**(6), 458–462 (2006)
9. Zhao, J., Zhao, Q., Chen, L., et al.: Effect of different surface treatments on corrosion behavior of 300M steel in Qingdao marine atmosphere. J. Chin. Soc. Corros. Prot. **39**(6), 504–510 (2019)
10. Jimmy, X., Guangling, S., Andriej, A.: Experimental measurement and computer simulation of galvanic corrosion of magnesium coupled to steel. Adv. Eng. Mater. **9**(2), 65–74 (2007)

Mechanical Behavior of Cerium Oxide-Reinforced Copper Surface Composites: A Comprehensive Study and Optimization

Dinesh Kumar[1], Satnam Singh[2(✉)], and Surjit Angra[2]

[1] Department of Mechanical Engineering, Maharishi Markandeshwar (Deemed to be University), Mullana-Ambala, India
dinesh_kumar@mmumullana.org

[2] Department of Mechanical Engineering, National Institute of Technology Kurukshetra, Kurukshetra, India
satnamsingh@nitkkr.ac.in

Abstract. The present investigation involved the application of the friction stir technique (FSP) on copper metal matrix composites (CMCs) that were reinforced with cerium oxide nanoparticles. The CMCs possess a range of advantageous attributes, including their mobility, power, potential for expansion, and inherent intelligence. The research also focuses on the optimization of parameters for friction stir processed surface composites (FSP-ed) utilizing the Taguchi method. This study examined the ideal combination of process parameters and its subsequent impact on hardness and tensile strength, eventually resulting in the optimal solution. The material has been shown to possess an optimum tensile strength of 440 MPa and hardness of 104.043 VHN. The optimal parameters for the experiment were determined to be a rotation speed of 1100 revolutions per minute (rpm), a transverse speed of 30 mm per minute (mm/min), and a total of 3 passes in each direction.

Keywords: Copper · cerium oxide · hardness · tensile strength · Taguchi method · friction stir processing

1 Introduction

Access to innovative materials is crucial for every area. Composites are a viable alternative for meeting these requirements. Composites are materials composed of many phases, resulting from the combination of two or more constituent components [1]. The two distinct processes in this context are matrix formation and reinforcement. The process of generating multi-phase materials involves the deliberate introduction of artificial additives into a foundational material, as opposed to their spontaneous occurrence. Composite materials are formed by the transformation of many substances. Due to their inherent differences, these two phases demonstrate a wide range of innovative or enhanced characteristics. Due to their outstanding characteristics, composites are increasingly being utilized in several areas, such as the automotive, mechanical, and aviation industries.

© The Author(s), under exclusive license to Springer Nature Singapore Pte Ltd. 2025
J. Mo and P. Chutima (Eds.): ICMSC 2024, LNME, pp. 13–20, 2025.
https://doi.org/10.1007/978-981-96-5380-5_2

Engineers frequently employ the practice of amalgamating a minimum of two disparate elements in order to enhance the physical, electrical, and mechanical properties of a composite [2]. Composites typically consist of a single continuous component, referred to as a matrix, and an uneven phase known as reinforcement. Matrix materials encompass a range of substances such as metals, ceramics, textiles, and polymers. On the other hand, reinforcements are comprised of rigid particles that are employed to enhance the structural integrity of the matrix material. Graphite (Gr) and Molybdenum disulfide (MoS2) are utilized in the construction of components that exhibit exceptional wear resistance. This is achieved by using stronger reinforcements to enhance desirable characteristics, such as the incorporation of softer lubricating reinforcements [3]. Composites are a distinct category of sophisticated engineering materials that possess the unique characteristic of including discrete reinforcement or fillers with a continuous matrix. By appropriately selecting the matrix and reinforcing components, the composite material can exhibit a diverse array of properties. Carbon fiber composites and fiberglass composites share a similarity in that they both consist of carbon-based materials for both the reinforcing and matrix components [4]. The incorporation of polymer as the matrix in these fillers is employed to enhance the mechanical properties of the composite, owing to the inherent brittleness of the fillers. This phenomenon arises due to the high modulus and strength exhibited by the reinforcements. Recent studies have demonstrated that the incorporation of ceramic material as a reinforcing agent in the base material enhances the physical and mechanical properties of the resulting composites [5]. This advancement presents novel prospects for the utilization of these composites in sectors such as electronic packaging and related fields [6–8]. The high thermal and electrical conductivity of copper (Cu) has rendered it valuable in several applications, such as electronic packaging, welding electrodes, and thermal connections. However, it is well acknowledged that pure copper does not possess the desired mechanical qualities, such as tensile strength, yield strength, creep resistance, and hardness. The aforementioned characteristics of the resulting metal matrix composites (MMCs) can be enhanced through the incorporation of reinforcements composed of ceramics, carbide, or oxides inside the copper matrix [9].

2 Materials and Method

2.1 Matrix Material

The focus of this study is on particle-reinforced composites, which involve the use of discrete reinforcements. These composites have been chosen due to their accessibility, affordability, ease of distribution within the matrix, and relatively uniform dispersion throughout the matrix. The selection of reinforcing materials is informed by the aims and uses of the composite [10]. In recent years, there has been a growing interest in the use of copper matrix composites within the aerospace and automotive industries. This is mostly attributed to their enhanced performance when compared to pure copper and its alloys. In this study, copper is utilized as the foundational alloy. The essential elements of the basic alloy are presented in Table 1.

Table 1. Chemical composition of pure copper.

Elements	Wt.%	Elements	Wt.%
Si	0.00073	**Sb**	0.0005
P	0.0021	**Co**	0.0002
Ti	0.0002	**Ni**	0.0038
Mn	0.0251	**Nb**	0.0002
V	0.0002	**Cu**	99.96

2.2 Reinforcement Material

The cerium oxide is used as filler in the pure copper during friction stir processing as mentioned in Table 2.

Table 2. Specifications of cerium oxide.

Reinforcement	C_eO_2
Type	Nano Powder
Melting Point	2300 °C
Particle Size	3 nm
Density (g/cm^3)	7.22

2.3 Friction Stirs Processing Procedure

The fundamental material utilized in this work consisted of pure copper plates with dimensions of 160 × 50 × 6 mm^3. Within these plates, a square measuring 1 × 1 mm was precisely incised along the axis of rotation. The crude cerium oxide underwent a processing time of 40 min subsequent to its acquisition and pulverization in a treated steel ball mill operating at a velocity of 140 m/s. The intended dimensions of the powder are typically 6 m. The process of crushing the notch into cerium oxide powder facilitates its handling and manipulation. The CNC processing machine developed by SIEMENS, which has a stacking limit of 1.5 tonnes, underwent modifications to accept the tube-shaped FSP device made of H13 steel. The FSP device has a shoulder width of 19 mm, a pin diameter of 4.7 mm, and a pin length of 4.7 mm. Pin-less mechanisms were employed to effectively seal the furrow and prevent the dispersion of reinforcing particles from the atmosphere. The cases investigated the parameters of instrument rotational speed (900–1100 rpm), translational speed (20–40 mm/min), passes per treatment (1–3), and slant point of the device. The establishment of these boundaries is achieved by conducting trials that effectively eliminate the potential for deformation. A total of nine distinct instances were considered, and it is customary for notable inaccuracies to arise during each subsequent modification. The study also included an examination of copper and a sample that was handled, non-reinforced, and subjected to FSP. The schematic representation of the FSP setup is depicted in Fig. 1.

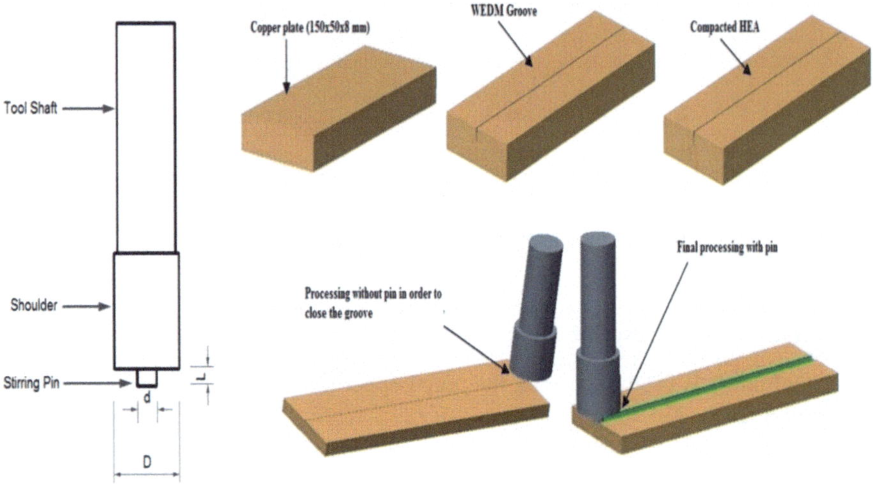

Fig. 1. Schematic diagram of friction stirs process.

3 Results and Discussions

3.1 Experimentation Outcomes

Following the application of friction stir processing (FSP) on copper plates, an examination was conducted on the FSP-treated specimens to assess their hardness and tensile strength. The results are presented in the table provided below (Table 3):

Table 3. Experimental values of hardness and tensile strength.

S. No.	Rotational Speed (rpm)	Transverse Speed (mm/min)	No. of Passes	Experimental Values	
				Hardness (VHN)	Tensile Strength (MPa)
1	900	20	1	73.40	227.42
2	900	30	2	76.78	224.25
3	900	40	3	63.82	198.12
4	1000	20	2	96.32	302.23
5	1000	30	3	106.21	326.27
6	1000	40	1	84.72	244.38
7	1100	20	3	144.84	441.74
8	1100	30	1	141.18	439.64
9	1100	40	2	87.87	263.34

3.2 Taguchi Method

The Taguchi technique is a proficient problem-solving methodology that has the potential to enhance the performance of products, processes, designs, and systems, while concurrently minimizing the duration and expenses associated with experimentation. The application of this technique, which combines experimental design theory with the concept of the quality loss function, has proven advantageous in enhancing the resilience of processes and products, as well as addressing other intricate issues encountered in the manufacturing industry. In addition, the methodology discerns the variables that exert the most significant influence on the result [11]. The ideal process parameters proposed by Taguchi exhibit robustness in the face of variations in ambient conditions and other sources of background noise. Increasing the number of process factors leads to a greater need for experimentation. In order to address this issue, the Taguchi technique utilizes an orthogonal array design to do a thorough examination of all potential process factors, while utilizing a minimal sample size. Taguchi categorizes the analysis of the Signal-to-Noise (S/N) ratio into three quality attributes: lower, larger, and nominal ratios. By employing S/N analysis, it is possible to ascertain the signal-to-noise ratio for each process variable. A higher signal-to-noise ratio (S/N) is indicative of improved quality features in all aspects. Consequently, the optimal value for the relevant process parameter is associated with the highest signal-to-noise ratio. An analysis of variance (ANOVA) can also be conducted to ascertain the statistical significance of the differences among components. The statistical importance of the process parameter for each quality attribute has been examined in previous studies [12–14].

3.3 ANOVA for Hardness

An analysis of variance (ANOVA) was conducted to examine the significance and impact of each process parameter on the outcome, with a focus on their relative importance and contribution. The investigation was carried out at a confidence level of 95% [15]. Table 4 displays the ANOVA results found for hardness. The findings of the study indicate that the dominant factor influencing hardness is the rotating speed, accounting for 61.69% of the observed variation. Subsequently, the transverse speed contributes 26.68% towards

Table 4. Analysis of variance for Hardness.

Source	DF	Adj SS	Adj MS	F-Value	P-Value	Contribution %
Rotational Speed (rpm)	2	3909.3	1954.6	16.57	0.057	61.69
Transverse Speed (mm/min)	2	1690.6	845.3	7.17	0.122	26.68
No. of Passes	2	501.5	250.7	2.13	0.320	7.91
Error	2	235.9	117.9			3.72
Total	8	6337.2				
		$R^2 = 96.28\%$			$R^2(adj) = 85.11\%$	

the hardness values, while the number of passes has a relatively less impact at 7.91%. These results were obtained in the pursuit of identifying optimal values for hardness. The results indicating the relevance of each component are presented in Table 4.

3.4 Analysis of Hardness and Tensile Strength

Figure 2 illustrates the Vickers hardness and tensile strength values of the copper-based surface composites that were manufactured. An increase in hardness and tensile strength was found following the introduction of hard reinforcement particles and their uniform dispersion within the matrix material [16–18]. Table 5 indicates each component of ANOVA analysis for tensile strength.

Table 5. Analysis of variance for Tensile strength.

Source	DF	Adj SS	Adj MS	F-Value	P-Value	Contribution %
Rotational Speed (rpm)	2	41032	20516	14.05	0.066	62.05
Transverse Speed (mm/min)	2	16805	8402	5.76	0.148	25.41
No. of Passes	2	5370	2685	1.84	0.352	8.12
Error	2	2919	1460			4.41
Total	8	66126				
		$R^2 = 95.59\%$			$R^2(adj) = 82.34\%$	

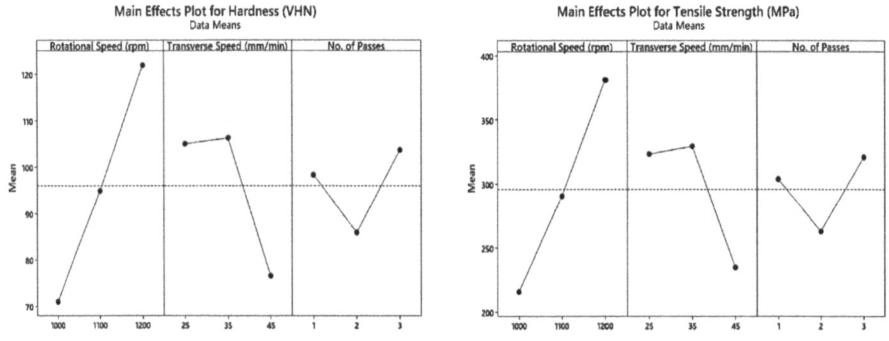

Fig. 2. Effect of process parameters on Hardness and Tensile strength

Regression Equations:

Hardness (VHN) = $RS_3 + TS_2 + NP_3 - 2H$

RS_3 is mean of hardness value at third level of rotational speed (rpm),

TS_2 is mean of hardness value at second level of transverse speed (mm/min),
NP_3 is mean of hardness value at third level of number of passes,
H, is overall mean of hardness (VHN).
Hardness (VHN) = 122.05 + 106.28 + 103.71 – 2 × 95.99 = 140.043 VHN.
Tensile Strength (MPa) = $RS_3 + TS_2 + NP_3 - 2Ts$
RS_3 is mean of tensile strength value at third level of rotational speed (rpm),
TS_2 is mean of tensile strength value at second level of transverse speed (mm/min),
NP_3 is mean of tensile strength value at third level of number of passes,
H, is overall mean of tensile strength (MPa).
Tensile Strength (MPa) = 381.20 + 329.58 + 321.14 – 2 × 295.98 = 440 MPa.

4 Conclusions

The aforementioned results have been derived by the implementation of experiments, employing the Taguchi technique, and the subsequent compilation of the acquired data.

The creation of surface composites of copper reinforced with cerium oxide was successfully achieved by employing friction stir processing.

The model's R-squared values are 96.28% and 95.59%, indicating a strong fit. Additionally, regression equations have been developed to accurately predict hardness and tensile strength based on any combination of input factors.

The analysis of variance yielded findings indicating that the optimal values for hardness (140.043 VHN) and tensile strength (440 MPa) were obtained by employing the following parameters: a rotational speed of 1100 rpm, a transverse speed of 30, and a third number of passes.

References

1. Kumar, D., Angra, S., Singh, S.: Mechanical properties and wear behaviour of stir cast aluminum metal matrix composite: a review. Int. J. Eng. Trans. A Basics **35**(4), 794–801 (2022). https://doi.org/10.5829/IJE.2022.35.04A.19
2. Singh, S., Angra, S.: Flexural and impact properties of stainless steel based glass fibre reinforced fibre metal laminate under hygrothermal conditioning, vol. 31, no. 1, pp. 164–172 (2018)
3. Kumar, D., Singh, S., Angra, S.: Effect of reinforcements on mechanical and tribological behavior of magnesium-based composites: a review. https://doi.org/10.18149/MPM.xxx2022
4. Sanesh, K., Sunder, S.S., Radhika, N.: Effect of reinforcement content on the adhesive wear behavior of Cu10Sn5Ni / Si 3 N 4 composites produced by stir casting **24**(9), 1052–1060 (2017). https://doi.org/10.1007/s12613-017-1495-1
5. Khosravi, J., Kazem, M., Givi, B.: Microstructural, mechanical , and thermophysical characterization of Cu / WC composite layers fabricated via friction stir processing, 1087–1096 (2014). https://doi.org/10.1007/s00170-014-6050-x
6. Khiyavi, B.A., Aghchai, A.J., Arbabtafti, M., Kazem, M., Givi, B.: Effect of friction stir processing on mechanical properties of surface composite of Cu reinforced with Cr particles **829**, 851–856 (2014). https://doi.org/10.4028/www.scientific.net/AMR.829.851
7. Sathish, T., Chandramohan, D., Vijayan, V., Sebastian, P.J.: Investigation on microstructural and mechanical properties of Cu reinforced with Sic composites prepared by microwave sintering process **009**, 5–9 (2019)

8. Stalin, B., et al.: Effect of titanium diboride ceramic particles on mechanical and wear behaviour of Cu-10 wt % W alloy composites processed by P / M route. Vacuum **184**, 109895 (2021). https://doi.org/10.1016/j.vacuum.2020.109895
9. Aktar Zahid Sohag, M., Gupta, P., Kondal, N., Kumar, D., Singh, N., Jamwal, A.: Effect of ceramic reinforcement on the microstructural, mechanical and tribological behavior of Al-Cu alloy metal matrix composite. Mater. Today Proc. **21**(xxxx), 1407–1411 (2020). https://doi.org/10.1016/j.matpr.2019.08.179
10. Jeon, C., et al.: Material properties of graphene / aluminum metal matrix composites fabricated by friction stir processing **15**(6), 1235–1239 (2014). https://doi.org/10.1007/s12541-014-0462-2
11. Xiong, Y., Wang, W., Shi, Y., Jiang, R.: Investigation on surface roughness, residual stress and fatigue property of milling in-situ TiB 2 / 7050Al metal matrix composites. Chin. J. Aeronaut. **34**(4), 451–464 (2021). https://doi.org/10.1016/j.cja.2020.08.046
12. Varalakshmi, K., Babu, A.J.: Analysis of dry sliding wear behaviour of Al 6061-coconut shell ash metal matrix composites using stir casting. Appl. Eng. Lett. J. Eng. Appl. Sci. **4**(2), 55–65 (2019). https://doi.org/10.18485/aeletters.2019.4.2.3
13. Khatkar, S.K., Verma, R., Kharb, S.S., Thakur, A., Sharma, R.: Optimization and effect of reinforcements on the sliding wear behavior of self-lubricating AZ91D-SiC-Gr hybrid composites, pp. 1461–1473 (2021)
14. Kumar, N., Bharti, A.: Optimization of powder metallurgy process parameters to enhance the mechanical properties of AZ91 magnesium alloy **48**(3), 315–327 (2022). https://doi.org/10.18149/MPM.4832022
15. Marakini, V., Pai, S., Bhat, A., Bangera, S.: Surface integrity optimization in high speed milling of AZ91 magnesium alloy using TOPSIS considering vibration signals. Mater. Today Proc. **52**, 802–809 (2022). https://doi.org/10.1016/j.matpr.2021.10.154
16. Liu, S., Wang, Y., Muthuramalingam, T., Anbuchezhiyan, G.: Effect of B4C and MOS2 reinforcement on micro structure and wear properties of aluminum hybrid composite for automotive applications. Compos. Part B Eng. **176**, 107329 (2019). https://doi.org/10.1016/j.compositesb.2019.107329
17. Kumar, P., Mahamani, A., Durga Prasad, B.: Experimental analysis of WAAM of ER4043-ALN composite for the automotive industry. J. Aeronaut. Mater. **42**(05), 142–151 (2022)
18. Rathi, P., Kumar, M., Sharma, R., Ritu: Analysis of hydrothermal synthesis of SnO2 – Al2O3 nanocomposite. J. Aeronaut. Mater. **42**(05), 152–158 (2022)

Decreasing Frequency and Electrochemical Potential as Means to Estimate Early Corrosion Fatigue Failure

Marcus Wolf[1,2] and Anja Pfennig[1(✉)]

[1] School of Engineering and Life Sciences, HTW Berlin, Berlin, Germany
anja.pfennig@htw-berlin.de

[2] BAM Federal Institute Materials Research and Testing, Berlin, Germany
marcus.wolf@bam.de

Abstract. In geothermal energy production facilities, the corrosion resistance and therefore the longevity of engineering steels is a vital factor of costs. DSS X2CrNiMoN22-5-3, a standard stainless duplex steel (AISI 2205) promises to withstand both, highly aggressive corrosive environment and mechanical load. Its electrochemical potential was monitored in-situ during corrosion fatigue experiments using a special corrosion fatigue test set-up and corrosion chamber (33 Hz, 369 K at a theoretical rate of saline water flow $\omega 0 = 1.7 \times 10^{\wedge} - 3$ m/s) allowing for in-situ monitoring of temperature, potential, frequency and pH of the surrounding geothermal brine. The agreement of the typical decline of the electrochemical potential with initiation as well as increasing propagation velocity of the crack as a function of the absolute potential are demonstrated - possibly enabling early prediction of failure.

Keywords: corrosion fatigue · frequency drop · duplex steel · electrochemical potential · corrosion

1 Introduction

Future applications in geothermal energy extraction, such as delivery pumps and components for conveying and filtration systems, present significant challenges for engineering materials. These materials must meet new demands, as they are subjected to mechanical loads while being exposed to highly corrosive environments. Corrosion processes are strongly influenced by mechanical stress, especially in relation to factors like cyclic loading frequency, temperature, and the concentration of chlorides in the corrosive media [1]. Little chromium content [2] and the occurrence of chlorides [3], as well as hydrogen sulfide (H_2S) [4], and carbonic dioxide (CO_2) [5] all accelerate corrosion fatigue in high-alloy steels. Increased temperatures, heavier mechanical loads, and minor pH levels further diminish the endurance limit, exacerbating corrosion fatigue effects in these materials [6]. However, higher chromium content and internal compressive surface stresses can enhance the endurance limit and improve corrosion resistance. Standard

© The Author(s), under exclusive license to Springer Nature Singapore Pte Ltd. 2025
J. Mo and P. Chutima (Eds.): ICMSC 2024, LNME, pp. 21–28, 2025.
https://doi.org/10.1007/978-981-96-5380-5_3

stainless duplex steel (DSS) X2CrNiMoN22-5-3 (AISI 2205), commonly used in various branches within industry [7], is known for its outstanding mechanical strength [7], opposition to corrosion cracking at induced stress (SCC) [8], and good corrosion resistance [9].

In a non-corrosive environment, the fatigue limit of SDD standard stainless duplex steel X2CrNiMoN22-5-3 is approximately 485 N/mm^2 (Pf = 50%, push-pull) after ten million cycles of testing [10–13]. When subjected to corrosion fatigue in saline brine from the Northgerman Geothermal Basin at 98 °C, with stress amplitudes extending from 175 to 325 N/mm^2 [10–13], the material endured a maximum of 9.2 × one million cycles at a stress amplitude of 240 N/mm^2 under free corrosion potential conditions (Fig. 1, left) [10–13]. The mechanical and corrosion behaviour is determined by surface quality [14]. Even surfaces as well as intrinsic plastic deformations and compressive stress at surfaces lead to high endurance limits [15]. At stress amplitudes above 275 N/mm^2 technical surfaces are superior over polished surfaces revealing more cycles to failure (Fig. 1, right) [10]. Main surface and pit precipitates consist of either iron carbonates or iron hydroxides (siderite $FeCO_3$, goethite FeOOH) independent of the external roughness [10, 11]. The surface state shows no significant impact on the consistent failure mechanism [11, 14] due to the relatively narrow scatter series (surfaces with technical finish TN = 1 / 1.35, surfaces with polished finish: TN = 1 / 1.95) of X2CrNiMoN22-5-3. Both mechanism, early pit formation as well as locally occurring depassivation in coexistence with early mechanical degradation [11] could lead to crack initiation.

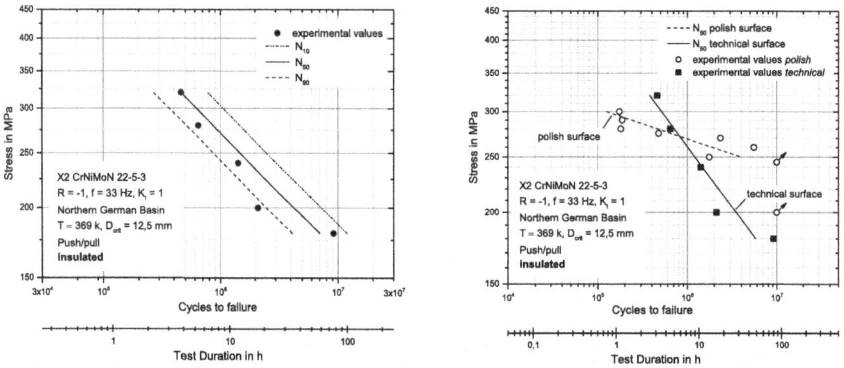

Fig. 1. Stress-Number to Failure curves for X2CrNiMoN22-5-3. Left: results obtained from an shielded experimental system [10, 11]. Right: evaluation of corrosion fatigue performance based on different surface structures. Technical surface: number of cycles to failure: P50% at stress amplitude 300 N/mm^2: 5 × 10^5 (k = 19.006). Polished surfaces: P50% at stress amplitude: 300 N/mm^2: 1.5 × 10^5 (k = 8.78) [11, 12].

This paper introduces the possibility to predict failure of corrosion fatigue test specimen in deep geothermal surroundings by in-situ monitoring of the electrochemical potential.

2 Experimental Procedure

Corrosion fatigue samples (surface roughness (polished): 0.58 μm–2.01 μm, average 1.59 μm) were made of stainless duplex steel X2CrNiMoN22-5-3 [11] (Table 1).

Table 1. Chemical arrangement of X2CrNiMoN22-5-3 (yield strength: 672 N/mm^2 and tensile strength of 854 N/mm^2).

phases	Carbon	Silica	Manganese	Phosphor	Sulfur	Chromium	Molybdenum	Nickel
$\alpha\&\gamma^b$	0.023	0.48	1.83	0.024	0.008	22.53	2.92	5.64
α^a	0,02	0.55	1.59	–	–	24.31	3.62	3.81
γ^a	0,03	0.47	1.99	–	–	20.69	2.17	6.54

[a]SPEKTROLAB M (Spektro) spark emission spectrometry
[b]JXA8900-RL Electron Probe Microanalyzer

The corrosive medium fluid is synthetic geothermal brine with pH6 (Northgerman basin [16]), as shown in Table 2).

Table 2. Chemical arrangement of the main components of the Northgerman basin.

	Sodium chloride	Hydrous calcium chloride	Potassium chloride	Magnesium chloride
g/L	98.22	207.24	5.93	4.18

The corrosion fatigue test procedure followed the standards of DIN EN ISO 11782-1 and the recommendations from FKM Research Issue 217 and were conducted using a Schenck-Erlinger Puls PPV testing machine operating at a frequency of 33 Hz. The test environment was maintained at 369 K, and the corrosion medium flowed at a rate of $V^* = 2.5 \times 10^{-6}$ m^3/s, achieving a given flow rate of $\omega_0 = 1.7 \times 10^{-3}$ m/s at the specimen's critical section. A corrosion compartment was securely fixed directly onto the sampling, allowing continuous monitoring during the mechanical tests. Data collected included electrochemical potential, temperature, and pH levels. For electrochemical measurements, a shock-resistant Ag/AgCl electrode was positioned in a PTFE sleeve (Fig. 2, middle right) [6, 10, 11].

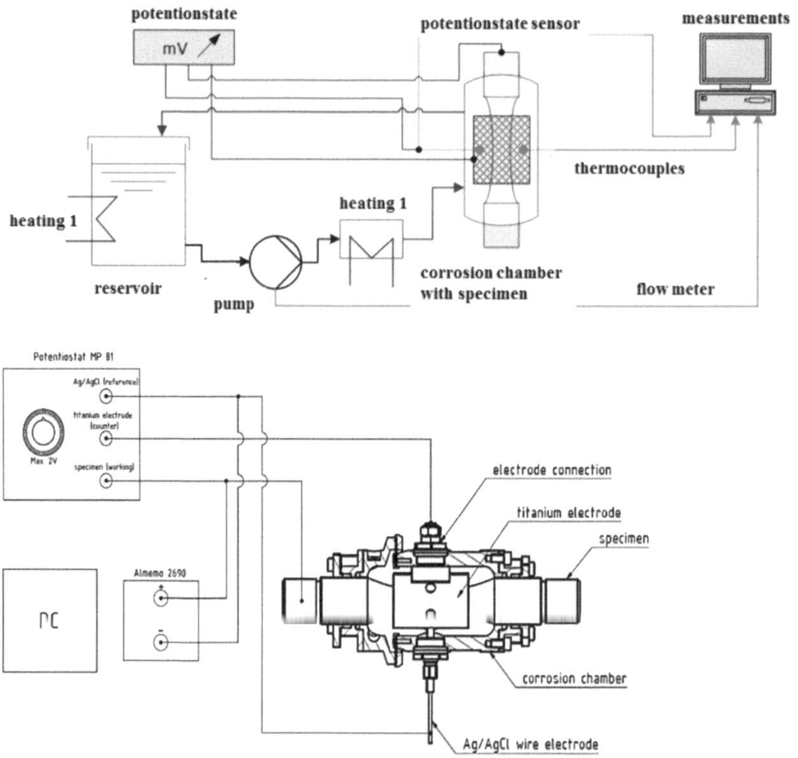

Fig. 2. Schematic set-up of corrosion fatigue experiments and collection of electrochemical data. The set-up covering multiple aspects was pronounced earlier [6, 10–13].

3 Results and Discussion

Samples are subjected to 72 h of testing, equivalent to 10^7 cycles in the corrosion chamber, where the potential remains unchanged compared to the initial values, indicating the setup does not affect electrochemical measurements. Significant potential changes correlate with cracking and failure during fatigue testing, highlighting the importance of mechanical load and crack propagation in potential variations (Fig. 3).

During fatigue testing, the potential measured against an Ag/AgCl reference electrode, was approximately 100 mV. Throughout the tests, there was a change in potential of about 100 mV from start to finish (Fig. 4). For specimens that did not experience failure, the potential remained relatively stable.

Since there is no indication that the surface condition influences the crack trail or the microstructure surrounding the crack site, the most effective explanation for the final failure is residual stress [16]. Therefore, it is necessary to predict early failure or failure on-set before final crack growth occurs. This is possible via close monitoring of the electrochemical potential. The onset of drop in electrochemical potential as associated with a change in frequency and therefore crack commencement.

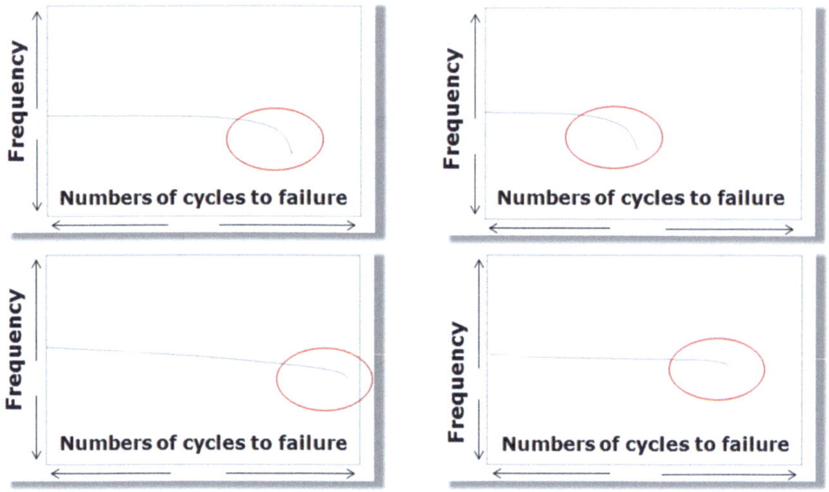

Fig. 3. Examples of electrochemical potential behaviour (Ag/AgCl electrode) during fatigue tests.

The drop of the potential at free corrosion potential (Fig. 4), that was monitored and assigned to failure on-set is typical for all corrosion fatigue specimen of standard stainless duplex steel X2CrNiMoN22-5-3 regarding both, absolute value of the potential and angle of inclination. Generally, the absolute potential of precision turned specimen exceed those of polished specimen by 50 mV. The time span in which the electrochemical potential declines is averagely 15 min. Figure 5 demonstrates free potential of an example specimen (N = 5.55·106, Sa = 260 MPa) shortly before failure. It remains stable for approximately 30 h. Subsequently typical fluctuations of the potential indicate initial local damage of the passivation layer. Because the voltage is low failure of the specimen did not occur immediately. Moreover, alternating reactions at the metal surface in contact with the geothermal brine result in passivation and depassivation predominantly impairing the steel. One might appoint the electrochemical fluctuations to a secondary flaw or micro crack. However, in this case only one crack passed through the specimen.

Crack propagation was estimated from the residual fatigue crack surface according to Wolf [12]. Crack propagation is evaluated in the area of crack depth where the crack propagation velocity is linear. Therefore, shortly before crack initiation the declination of the potential is proportional to the increase in area of fatigue crack area. Due to the high mean variance 10 measured data points of the potential were assigned to one average potential number. Decline of the potential is in good agreement with the crack commencement and estimated increasing crack propagation velocity.

Note, that this method to estimate crack propagation has to be evaluated critically because exact crack initiation may not be derived exactly from the potential graph. It may vary about many thousands of cycles to failure. Moreover, local corrosion phenomena are not considered in the model. Because all cracks are initiating from a corrosion pit which is dependent on the applied stress amplitude the determination of the crack depth would have to be reconsidered. However, the pit at approximately 5–10 μm is rather small and would account for only 0.8% of the crack depth and therefore might be neglected.

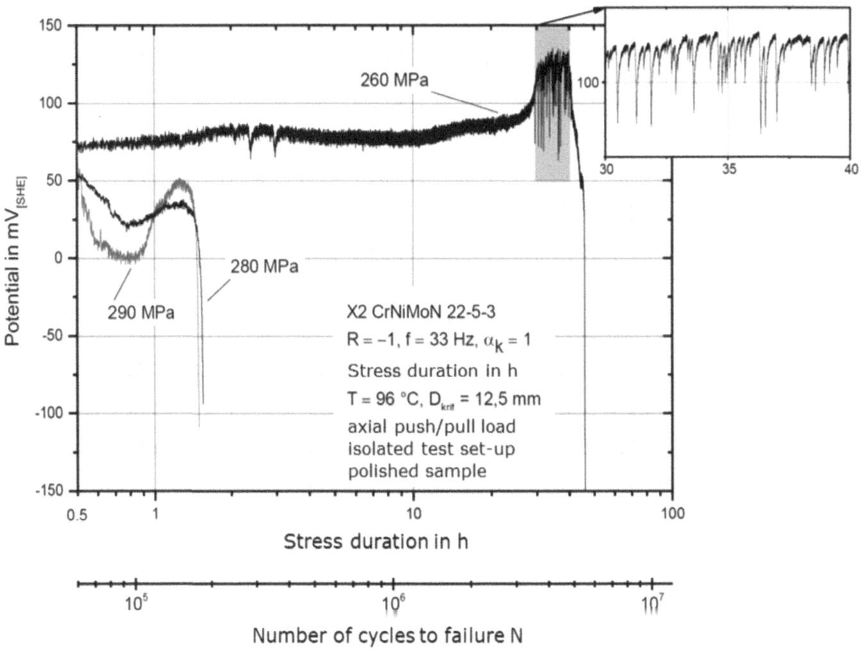

Fig. 4. Electrochemical potential of X2CrNiMoN22-5-3 monitored in the period of specimen failure under corrosion fatigue conditions at axial push/pull load.

The specific potential drop in the final experimental phase caused by crack propagation specific potential drops is now used to develop an algorithm to predict the specimen's failure. The aim was to achieve a time lead in comparison to the conventional test cancellation procedure where the drop in frequency has to be awaited. Because the specimen show supercritical crack propagation the determination of the onset of crack initiation would also allow extracting new information about the CF crack initiation procedure. In addition, early warning systems for geothermal application could be developed eventually. The algorithm is based on excruciating the measured potential numbers into ten period series with a capacity of ten single numbers each. By combining the splitting with mean value, slope and regression calculations the algorithm could be successfully initiated. The evaluation of the functionality is approved by fifteen potential functions. Up to now the employment of this algorithm showed a functionality level of 93.3% and resulted in a time lead up to 435 [13] seconds. The development of the algorithm is subject to future research activities.

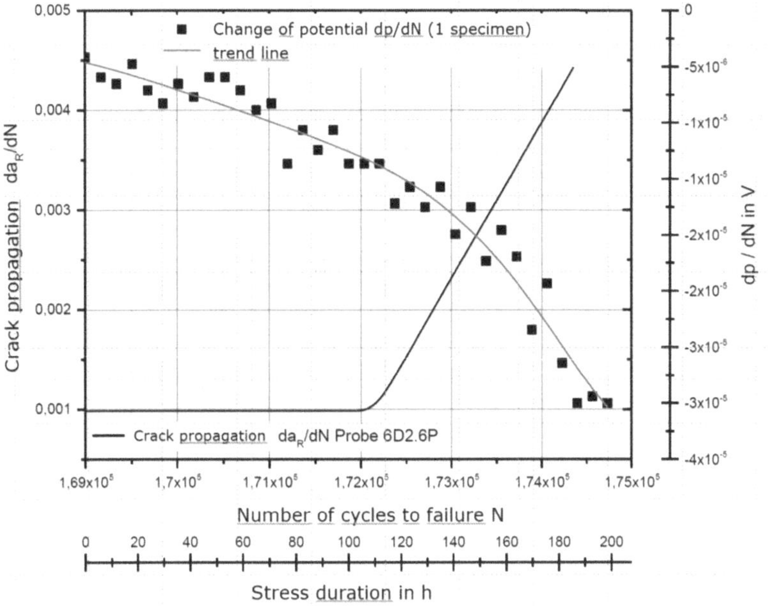

Fig. 5. Electrochemical potential of X2CrNiMoN22-5-3 as a function of crack propagation under corrosion fatigue conditions at axial push/pull load.

4 Summary

The prediction of early failure from the electrochemical potential of a steel exposed to highly corrosive environment and at the same time mechanical stress would reduce maintenance and expenses in geothermal energy production facilities. Therefore, the electrochemical potential of standard stainless duplex steel DSS X2CrNiMoN22-5-3 (AISI 2205) (N = 5.55·106, Sa = 260 MPa) was monitored in-situ during corrosion fatigue experiments. The typical drop of the potential at free corrosion potential was assigned to the failure on-set regarding both, absolute value of the potential and angle of inclination (averagely 15 min). The absolute potential of precision turned specimen exceeds those of polished specimen by 50 mV. Typical fluctuations of the potential indicate initial local damage of the passivation layer with passivation and depassivation reactions impairing the steel. Crack propagation is derived from the residual fatigue crack surface guaranteeing linear crack propagation velocity. The method has to be evaluated critically because firstly the crack initiation may not be derived exactly from the potential graph and secondly local corrosion phenomena are not considered in the model. However, the decline of the electrochemical potential is in good agreement with crack commencement and estimated increasing crack propagation velocity.

References

1. Thomas, J.P., Wei, R.P.: Corrosion fatigue crack growth of steels in aqueous solutions I: experimental results and modeling the effects of frequency and temperature. Mater. Sci. Eng. A **159**, 205–221 (1992)
2. Mu, L.J., Zhao, W.Z.: Investigation on carbon dioxide corrosion behaviors of 13Cr stainless steel in simulated strum water. Corros. Sci. **2**, 82–89 (2010)
3. Unigovski, Ya.B., Lothongkum, G., Gutman, E.M., Alush, D., Cohen, R.: Low-cycle fatigue behavior of 316L-type stain-less steel in chloride solutions. Corr. Sci. **51**, 3014–3120 (2009)
4. Holtam, C.M., Baxter, D.P., Ashcroft, I.A., Thomson, R.C.: Effect of crack depth on fatigue crack growth rates for a C-Mn pipeline steel in a sour environment. Int. J. Fatigue **32**, 288–296 (2010)
5. Chen, X., Yang, X., Zeng, M., Wang, H.: Influence of CO_2 partial pressure and flow rate on the corrosion behavior of N80 steel in 3.5% NaCl. Int. J. Electrochem. Sci. **18**(8) (2023)
6. Pfennig, A., Simkin, R.: Identifying the initial corrosion fatigue failure based on dropping electrochemical potential. Appl. Sci. **15**(1), 403 (2025)
7. Alvarez-Armas, I.: Duplex stainless steels: brief history and some recent alloys. Recent Pat. Mech. Eng. **1**, 51–57 (2008)
8. Prosek, T., et al.: Low-temperature stress corrosion cracking of austenitic and duplex stainless steels under chloride deposits. Corrosion **70** (2014)
9. Schultze, S., Göllner, J., Eick, K., Veit, P., Heyse, H.: Selektive Korrosion von Duplexstahl Teil 1: Aussagekraft herkömmlicher und neuartiger Methoden zur Untersuchung des Korrosionsverhaltens von Duplexstahl X2CrNiMoN22-5-3 unter besonderer Berücksichtigung der Mikrostruktur. Mater. Corros. **52/1**, 26–36 (2010)
10. Wolf, M., Pfennig, A.: Influence of geothermal environment on the corrosion fatigue behaviour of standard duplex stain-less steel X2CrNiMoN22-5-3. Model. Methods Struct. Anal. IOP Conf. Ser. J. Phys. Conf. Ser. (JPCS) **1425** (2020). Paper No. 012183
11. Pfennig, A., Wolf, M., Kranzmann, A.: Corrosion and corrosion fatigue of steels in downhole CCS environment—a summary. Processes **9**, 594 (2021)
12. Wolf, M.: Dissertation OVG Madeburg (2018)
13. Simkin, R.: Bachelor thesis 14.9.2015, HTW-Berlin
14. Zhang, W., Fang, K., Hua, Y., Wang, S., Wang, X.: Effect of machining-induced surface residual stress on initiation of stress corrosion cracking in 316 austenitic stainless steel. Corros. Sci. **108**, 173–184 (2016)
15. Abdulstaar, M., Mhaede, M., Wollmann, M., Wagner, L.: Investigating the effects of bulk and surface severe plastic deformation on the fatigue, corrosion behaviour and corrosion fatigue of AA5083. Surf. Coat. Technol. **254**, 244–251 (2014)
16. Forster, A., et al.: Reservoir characterization of a CO_2 storage aquifer: the Upper Triassic Stuttgart Formation in the Northeast German Basin. Mar. Pet. Geol. **27**, 2156–2172 (2010)

Mechanical Structure Design and Analysis

Virtual Spring-Assisted Constrained Method for Calculating Steady-State Balance Position of Flexible Rotors Supported by Hydrodynamic Bearings

Yong Guo[1(✉)], Shuyun Jiang[1], and Yuliang Lu[2]

[1] Southeast University, 2 Southeast Road, Nanjing 211189, China
guoyong@seu.edu.cn

[2] AVIC Nanjing Engineering Institute of Aircraft System,
33 Shuige Road, Nanjing 211106, China

Abstract. When combining the Newton-Raphson method with the transfer matrix method to solve the steady-state balance position of a flexible rotor supported by hydrodynamic bearings, one commonly faces sensitivity issues with initial values that make the iterative process difficult to converge. This paper proposes a novel method using virtual spring-assisted constraints, which effectively and conveniently addresses this problem. A specific computational example is provided to demonstrate that the proposed method achieves good convergence without relying on preliminary estimates of initial values. This paper also highlights the need for the virtual spring to have high stiffness. The iterative process described by the novel method conforms to actual physical behavior and is easy to understand intuitively and implement programmatically.

Keywords: Flexible Rotor · Equilibrium Position · Newton-Raphson Method

1 Introduction

In the analysis of electric spindle rotors, the rotors are often subjected to steady-state loads (such as constant cutting forces) and periodic loads (such as centrifugal forces caused by residual unbalance) [1, 2]. For rotors with high dynamic balance accuracy, the residual unbalance is small, and the steady-state loads dominate. According to the small perturbation assumption [3], the motion of the rotor can be decomposed into dominant steady-state quantities (i.e., steady-state displacements and static deformations) and relatively smaller perturbations (i.e., low-amplitude, high-frequency vibrations). Under steady-state displacements and static deformations, the positions occupied by various parts of the flexible rotor can be referred to as the steady-state balance positions.

For rotors supported by hydrodynamic bearings, the calculation of their steady-state balance positions is particularly important. The steady-state balance positions of flexible rotors determine the distribution of oil film thickness in the hydrodynamic bearings, which significantly influences the distribution of oil film pressure and, in turn, determines

© The Author(s), under exclusive license to Springer Nature Singapore Pte Ltd. 2025
J. Mo and P. Chutima (Eds.): ICMSC 2024, LNME, pp. 31–40, 2025.
https://doi.org/10.1007/978-981-96-5380-5_4

the static and dynamic characteristics of the bearings. Conversely, the support force and stiffness damping characteristics of the bearings profoundly affect the steady-state balance positions and dynamic characteristics of the flexible rotors [4]. Only by determining the steady-state balance positions of the rotors can accurate static and dynamic characteristics of the bearings be obtained, enabling the dynamic analysis of the flexible rotors in electric spindles.

The transfer matrix method is a widely used numerical method primarily applied for the dynamic characteristic analysis of rotor systems [1], including natural frequencies, modes, and responses. This method can also be extended for solving the static characteristics of rotor systems. Compared to the finite element method [5], the dimension of the matrices in the transfer matrix method does not increase with the increase in the number of degrees of freedom, making it more efficient for handling long and slender flexible rotor systems. The Newton-Raphson method is a method based on the function and its derivative for approximating the solution of nonlinear equations [6]. In the calculation of the steady-state balance positions of flexible rotors, the mechanical properties of the supporting hydrodynamic bearings are nonlinear. Through multiple iterations of the Newton-Raphson method, the exact solution can be gradually approached.

Combining the Newton-Raphson method with the transfer matrix method is an effective approach for addressing nonlinear static problems in flexible rotors. However, when applying this method to calculate the steady-state balance positions of flexible rotors supported by hydrodynamic bearings, the equations often fail to converge. This is because the Newton-Raphson method is a locally convergent method [7] whose convergence strongly depends on the estimation and selection of initial values. For a flexible rotor with five degrees of freedom, it is unrealistic to estimate the approximate displacements in all directions based on external loads, and often only zero displacement vector can be chosen as the initial value for the steady-state balance position. For hydrodynamic bearings, when there is no eccentricity, they do not produce a support force and cannot provide effective constraint on the rotor. As a result, the correction factor in the Newton-Raphson method becomes too large, causing the supported part of the rotor to move outside the bearing confine and leading to divergence in the calculations. In fact, due to the thin oil film in hydrodynamic bearings, the sensitivity of the steady-state balance position of flexible rotors to initial values is particularly pronounced under heavy loads. Thus, researchers often employ the finite element method to solve the static problems of flexible rotors supported by hydrodynamic bearings, and on this basis, they further use the transfer matrix method to address the dynamic problems of the rotors; such a combination of multiple methods increases the complexity of the entire analytical process.

This paper proposes an improved computational method, the virtual spring-assisted constrained Newton-Raphson method combined with the transfer matrix method. Using this method effectively solves the initial value sensitivity issue of the original method, allowing convergence in the calculation of the steady-state balance positions of flexible rotors regardless of the choice of initial values. Additionally, the method has fast convergence speed and is easy to program.

2 Method

2.1 Prohl Transfer Matrix Method

As shown in Fig. 1, the continuous flexible rotor is discretized along the axial direction into a system with concentrated masses and finite degrees of freedom. This process includes the equivalent concentration of mass and rotational inertia, the equivalent stiffness of the shaft, and the simplification of the rotor supports.

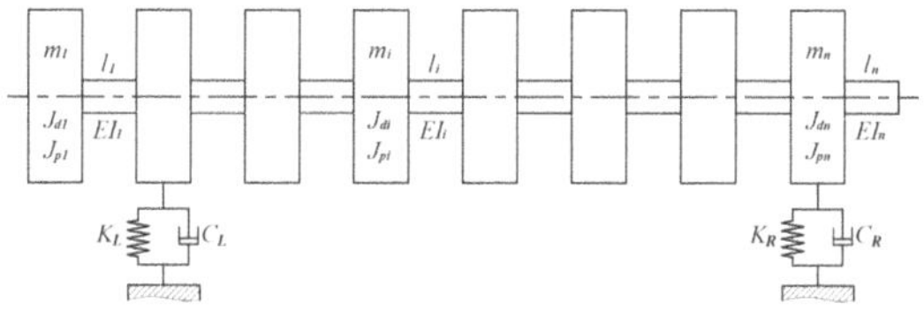

Fig. 1. Discretized flexible rotor model

Figure 2 shows the force model of the concentrated mass disk and shaft segment of unit i. Here, m and l represent the mass of the disk and the length of the shaft segment, respectively. Q and M represent the shear force and bending moment at the cross-section. Subscripts indicate the order of the unit, and superscripts indicate which side of the concentrated mass disk of unit i the force acts upon.

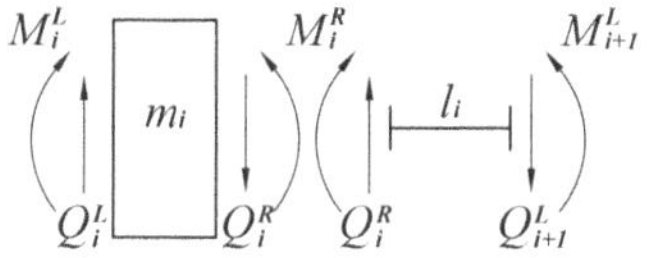

Fig. 2. Force model for disk and shaft segments

For the static problem of a five-degree-of-freedom flexible rotor, the state vector of the i-th unit can be represented as:

$$\mathbf{Z}_i^L = \left[(x, y, z, \theta_x, \theta_y, Q_x, Q_y, Q_z, M_x, M_y, 1)_i^L \right]^T \tag{1}$$

where the first five terms describe the displacements of the node of the unit, and the following five terms describe its force state. The transfer matrix of the i-th unit is:

$$
\boldsymbol{TM}_i = \begin{bmatrix}
1 & 0 & 0 & 0 & l_i & \frac{l_i^3}{6EI_i} & 0 & 0 & \frac{l_i^2}{2EI_i} & 0 & \frac{l_i^3 F_{xi}}{6EI_i} + \frac{l_i^2 T_{xi}}{2EI_i} \\
0 & 1 & 0 & -l_i & 0 & 0 & -\frac{l_i^3}{6EI_i} & 0 & 0 & \frac{l_i^2}{2EI_i} & -\frac{l_i^3 F_{yi}}{6EI_i} + \frac{l_i^2 T_{yi}}{2EI_i} \\
0 & 0 & 1 & 0 & 0 & 0 & 0 & 0 & 0 & 0 & 0 \\
0 & 0 & 0 & 1 & 0 & 0 & \frac{l_i^2}{2EI_i} & 0 & 0 & -\frac{l_i}{EI_i} & \frac{l_i^2 F_{yi}}{2EI_i} - \frac{l_i T_{yi}}{EI_i} \\
0 & 0 & 0 & 0 & 1 & \frac{l_i^2}{2EI_i} & 0 & 0 & \frac{l_i}{EI_i} & 0 & \frac{l_i^2 F_{xi}}{2EI_i} + \frac{l_i T_{xi}}{EI_i} \\
 & & & & & 1 & & & & & F_{xi} \\
 & & & & & 0 & 1 & & & & F_{yi} \\
 & & & & & 0 & 0 & 1 & & & F_{zi} \\
 & & & & & l_i & 0 & 0 & 1 & & l_i F_{xi} + T_{xi} \\
 & & & & & 0 & -l_i & 0 & 0 & 1 & -l_i F_{yi} + T_{yi} \\
 & & & & & 0 & 0 & 0 & 0 & 0 & 1
\end{bmatrix} \tag{2}
$$

where the force F and torque T reflect the support effect of the hydrodynamic bearings on the concentrated mass disks. For unsupported units, these terms are zero. After obtaining the transfer matrix for each unit, the state vector of the terminal node of the flexible rotor can be obtained from that of the initial node through successive steps:

$$
\boldsymbol{Z}_n^L = \prod_{i=1}^{n-1} (\boldsymbol{TM}_i) \cdot \boldsymbol{Z}_1^L \tag{3}
$$

Boundary conditions: To solve the static problem of a flexible rotor, appropriate boundary conditions should be introduced. For a common case where the front end of the spindle is subjected to a cutting force and the rear end is free, the force boundary condition at the terminal node is as follows:

$$
\left[(Q_x, Q_y, Q_z, M_x, M_y)_n^L\right]^T = (0, 0, 0, 0, 0)^T \tag{4}
$$

2.2 Virtual Spring-Assisted Constraint

When the initial values cannot be accurately estimated, it is typically assumed that the displacements in all directions at the initial node are zero. In the initial state without eccentricity, the hydrodynamic bearings do not provide any constraint on the flexible rotor. To avoid non-convergence in subsequent iterative calculations, additional virtual springs are introduced as auxiliary constraints at the disk nodes supported by hydrodynamic bearings.

As shown in Fig. 3(a), when the rotor is correctly positioned within the hydrodynamic bearing, constraints are applied to the rotor according to the mechanical characteristics of the bearing. As shown in Fig. 3(b), when the rotor position erroneously extends beyond the bearing confine, the virtual spring constraints are activated. Under the constraints of the virtual springs, after subsequent iterative calculations, the rotor position will

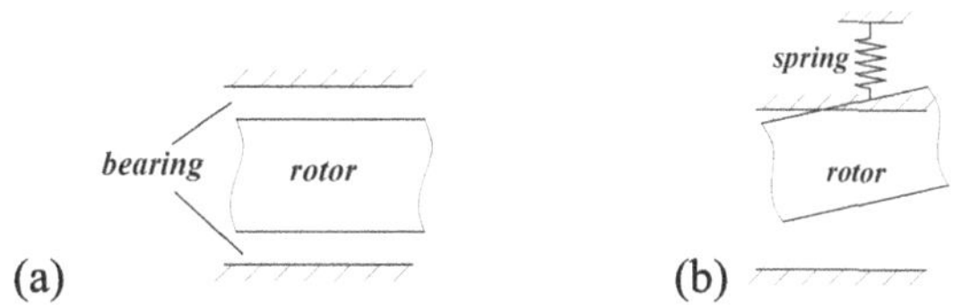

Fig. 3. Constraints of hydrodynamic bearing and virtual spring

inevitably return to within the hydrodynamic bearing, thereby obtaining effective support and constraints from the bearing.

The convenient criterion for determining whether the rotor is within the hydrodynamic bearing is the minimum film thickness of the bearing lubricant. If the value is positive, then the rotor is within the bearing, and the bearing constraints are invoked. Conversely, if the value is negative, the rotor extends beyond the bearing confine, and the virtual spring constraints are invoked. This constraint condition can be described as follows:

$$\mathbf{F} = \begin{cases} \mathbf{F}_{bearing}, h_{\min} > 0 \\ \mathbf{F}_{spring}, h_{\min} \leq 0 \end{cases} \tag{5}$$

2.3 Newton-Raphson Method

Construct the target vector based on the displacement state of the initial node of the flexible rotor:

$$\boldsymbol{q} = \left[x_1, y_1, z_1, \theta_{x1}, \theta_{y1}\right]^T \tag{6}$$

Given the displacement state $\boldsymbol{q}$ of the initial node, the state vector of the terminal node can be obtained through successive steps using (3), from which the force vector can be extracted. Therefore, the force vector at the terminal node can be written as a function $G(\boldsymbol{q})$ of $\boldsymbol{q}$. According to (4), the vector function $G(\boldsymbol{q})$ must satisfy the boundary condition:

$$G(\boldsymbol{q}) = (0, 0, 0, 0, 0)^T \tag{7}$$

The application of the Newton-Raphson method to solve the nonlinear Eq. (7) involves multiple iterations, during which the target vector is updated after each iteration:

$$\boldsymbol{q}^{k+1} = \boldsymbol{q}^k - G'(\boldsymbol{q}^k)^{-1} G(\boldsymbol{q}^k) \tag{8}$$

where $\boldsymbol{q}^k$ represents the vector obtained at the k-th iteration, and the equation is solved when $\boldsymbol{q}^k$ converges. $G'(\boldsymbol{q}^k)$ is its Jacobian matrix, expressed as:

$$G'(q) = \begin{bmatrix} \frac{\partial Q_{xn}}{\partial x} & \frac{\partial Q_{xn}}{\partial y} & \frac{\partial Q_{xn}}{\partial z} & \frac{\partial Q_{xn}}{\partial \theta_x} & \frac{\partial Q_{xn}}{\partial \theta_y} \\ \frac{\partial Q_{yn}}{\partial x} & \frac{\partial Q_{yn}}{\partial y} & \frac{\partial Q_{yn}}{\partial z} & \frac{\partial Q_{yn}}{\partial \theta_x} & \frac{\partial Q_{yn}}{\partial \theta_y} \\ \frac{\partial Q_{zn}}{\partial x} & \frac{\partial Q_{zn}}{\partial y} & \frac{\partial Q_{zn}}{\partial z} & \frac{\partial Q_{zn}}{\partial \theta_x} & \frac{\partial Q_{zn}}{\partial \theta_y} \\ \frac{\partial M_{xn}}{\partial x} & \frac{\partial M_{xn}}{\partial y} & \frac{\partial M_{xn}}{\partial z} & \frac{\partial M_{xn}}{\partial \theta_x} & \frac{\partial M_{xn}}{\partial \theta_y} \\ \frac{\partial M_{yn}}{\partial x} & \frac{\partial M_{yn}}{\partial y} & \frac{\partial M_{yn}}{\partial z} & \frac{\partial M_{yn}}{\partial \theta_x} & \frac{\partial M_{yn}}{\partial \theta_y} \end{bmatrix} \tag{9}$$

2.4 Computational Flowchart

The solution process involves the combined use of the Newton-Raphson method and the transfer matrix method, with the introduction of virtual spring auxiliary constraints to prevent non-convergence during the iteration process. The entire computational process is illustrated in Fig. 4:

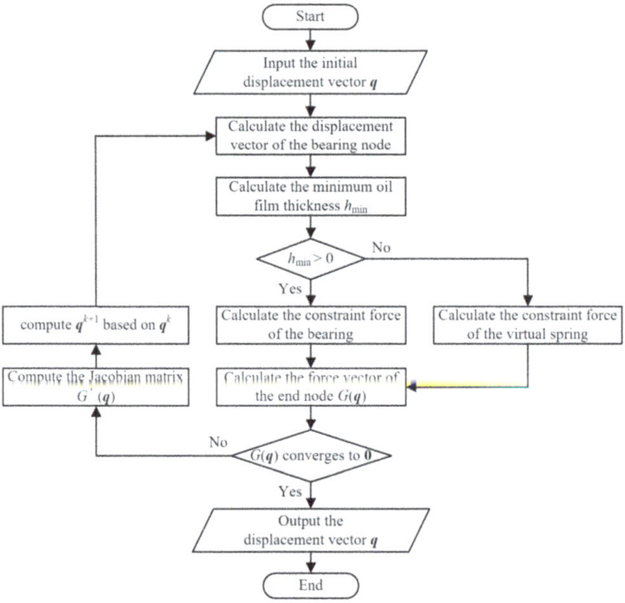

Fig. 4. Computational process flowchart

3 Computational Example and Analysis

Figure 5 shows a typical hydrodynamic bearing electric spindle, with the main dimensional parameters marked. The front end of the spindle is subjected to a constant cutting force, while the rear end is free. It is supported by a set of closed hydrodynamic thrust bearings and two journal bearings. For this example, the steady-state equilibrium position of the flexible rotor is solved using the virtual spring-assisted constrained Newton-Raphson method combined with the transfer matrix method, as proposed in this article. The main parameters of the rotor and the hydrodynamic bearings are given in Table 1.

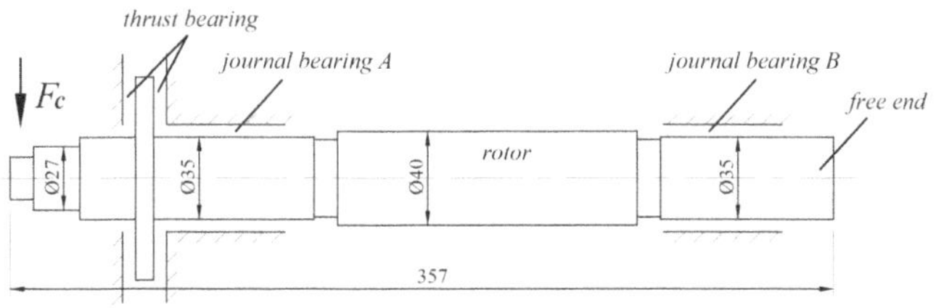

Fig. 5. Rotor and support of the electric spindle in this computational example

Table 1. Parameters of Rotor and Hydrodynamic Bearings

Parameters	Value
Rotor material's Young's modulus E	206 GPa
Hydrodynamic bearing clearance c	20 μm
Lubricant viscosity μ	8.6×10^{-3} Pa · s
Journal bearing radius R_j	17.5 mm
Journal bearing length L_j	60 mm
Thrust bearing outer radius R_o	43 mm
Thrust bearing inner radius R_i	17.5 mm

3.1 Convergence of the Novel Method

Figure 6 shows the relationship between the nodal displacements of journal bearings A and B and the iteration count at a rotational speed $\omega = 6000\,\text{rpm}$, cutting force $F_c = 300\,\text{N}$ (applied in the negative x-direction), and virtual spring stiffness $k_{xx} = k_{yy} = 10^4\,\text{N}/\mu\text{m}$. (a) to (d) illustrate the displacements of the nodes in various directions. It is evident that after eight iterations, the nodal displacements tend to stabilize, leading to the convergence of the steady-state equilibrium position.

In this example, the bearing clearance is 20 μm, and throughout the iterative process, the rotor remains within the confine of Bearing A. However, during the first and second iterations, the rotor evidently exceeds the constraint range of Bearing B. It is due to the action of the virtual spring that, by the third iteration, the rotor returns to within the limit of Bearing B, enabling the iterative process to proceed. This example demonstrates the good convergence of the novel method and overcomes the issue of sensitivity to initial values in computations.

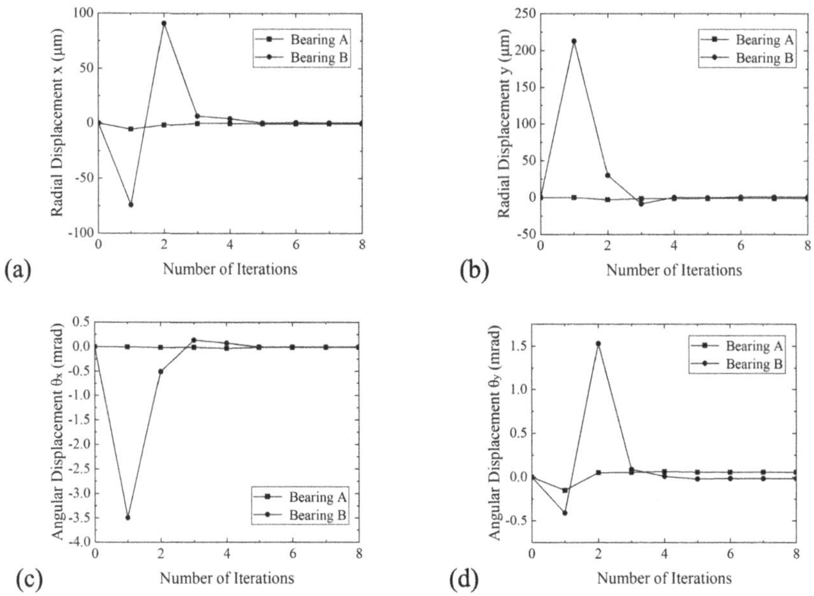

Fig. 6. Nodal displacements of bearing A and B versus iterations

3.2 Determination of Values for Virtual Spring Stiffness

Figure 7 illustrates the relationship between the eccentricity of Bearing B and the number of iterations for various virtual spring stiffness values. Due to fluctuations in the data range during the iteration process, a logarithmic scale is adopted for the eccentricity values to describe their variations. For Bearing A, as it consistently constrains the rotor throughout the iterative process, there is no need to activate the virtual spring for additional constraint. The figure demonstrates that once the selected virtual spring stiffness exceeds a specific threshold (100 N/μm in this example), the rate of computational convergence becomes highly consistent. For virtual springs with stiffness values of 10^2, 10^3, and 10^4 N/μm, the steady-state equilibrium positions of the rotor converge to the same value after eight iterations. As shown, the three curves nearly coincide.

However, for a virtual spring with a stiffness of 1 N/μm, the eccentricity of Bearing B converges to approximately 1.3. Clearly, when the bearing eccentricity $\varepsilon > 1$, the rotor is located outside the hydrodynamic bearing, indicating that the steady-state equilibrium position of the rotor lies beyond the bearing's constraint range, which is an obviously incorrect result. The cause of this error is that the selected virtual spring stiffness is too small, and the restoring force of the spring is insufficient to press the rotor back into the bearing when it deviates from the bearing's confine. Ultimately, the rotor's position converges to a balance between the external force and the virtual spring's constraint, which is not the desired outcome. Therefore, to obtain a correct convergence result, a higher stiffness should be chosen for the virtual spring, possibly related to the flexural rigidity of the rotor, the magnitude of the external load, and the bearing clearance. The specific value should be determined through numerical experiments. A convenient method is to judge based on the bearing eccentricity after the equation converges; if

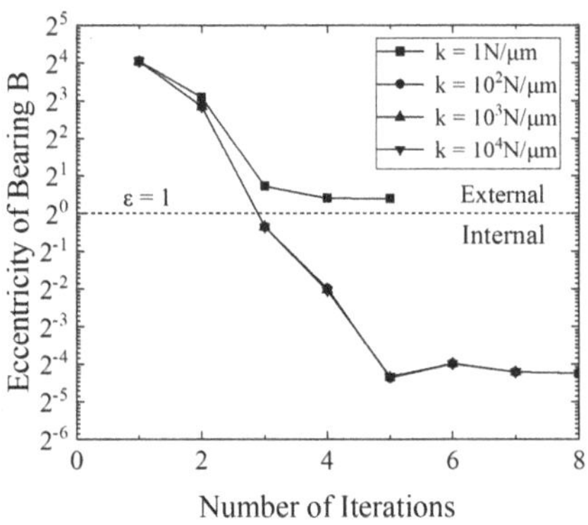

Fig. 7. Convergence of eccentricity under various spring stiffnesses

$\varepsilon \geq 1$, it indicates that the spring stiffness is too low and needs adjustment. Conversely, if $\varepsilon < 1$, the converged steady-state equilibrium position is the correct result we aim to achieve.

3.3 Steady-State Equilibrium Position of the Entire Rotor

Based on the novel method described herein, the steady-state equilibrium position of the flexible rotor as a whole is obtained, as depicted in Fig. 8.

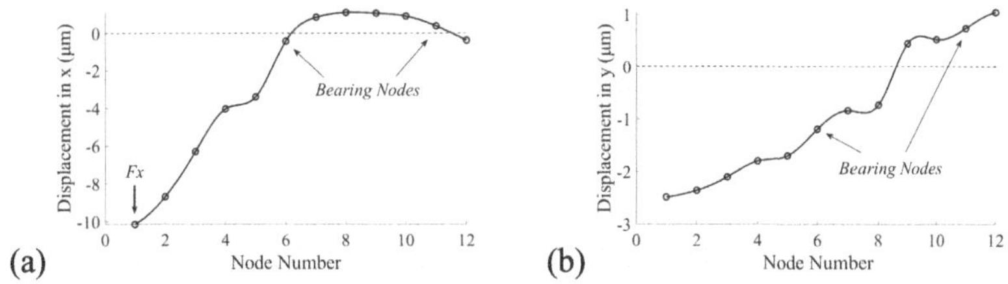

Fig. 8. Whole rotor steady-state equilibrium position

Due to the load applied at the initial node, the front end of the flexible rotor experiences a displacement of approximately 10 μm in the negative x-direction, as expected. At the nodes supported by the bearings, the rotor displacements are significantly smaller, which demonstrates the constraining effect of the bearings on the rotor. In addition, due to the coupled mechanical characteristics of the hydrodynamic bearings in multiple degrees of freedom, the load in the x-direction also induces displacement in the y-direction.

4 Conclusion

This paper presents a novel method for determining the steady-state equilibrium position of a flexible rotor supported by hydrodynamic bearings, which combines the Newton-Raphson method with the transfer matrix method and incorporates virtual spring auxiliary constraints. The specific procedure for solving the problem is outlined, and the effectiveness of this method is demonstrated through a specific computational example. Based on the results and analysis, the following key conclusions are drawn:

1. Utilizing virtual spring auxiliary constraints effectively prevents divergence in the iterative process for calculating the steady-state equilibrium position of a flexible rotor. This method exhibits excellent convergence properties, overcoming the limitation of local convergence inherent in the Newton-Raphson method and ensuring that the calculation process is independent of the initial estimate.
2. The stiffness of the virtual spring should not be excessively low, as this may lead to the steady-state equilibrium position of the flexible rotor converging to an incorrect solution that is constrained by the virtual spring rather than the actual hydrodynamic bearings. During the computation using this method, attention should be paid to whether the eccentricity of the bearings is less than 1 at convergence; if not, the stiffness of the virtual spring should be adjusted.
3. The proposed new method addresses the issue of the flexible rotor exceeding the bearing boundaries through physical simulation (via virtual spring constraint), ensuring that the iterative process is consistent with actual physical behavior and facilitating intuitive understanding and implementation in programming.

Acknowledgment. This work was financially supported by the National Science Foundation of China through Grant Nos. 12172088 and 51635004.

References

1. Huang, X., Xu, G., Jiang, S.Y.: Static characteristics of water-lubricated hydrodynamic spiral-groove journal and thrust bearings for motorized spindle. J. Tribol.-Trans. ASME **145**(12) (2023)
2. Wang, W.B., Song, P.Y., Yu, H.C., Zhang, G.Q.: Research on vibration amplitude of ultra-precision aerostatic motorized spindle under the combined action of rotor unbalance and hydrodynamic effect. Sensors **23**(1), 496 (2023)
3. Xu, C.D., Jiang, S.Y.: Dynamic analysis of a motorized spindle with externally pressurized air bearings. J. Vibr. Acoust.-Trans. ASME **137**(4) (2015)
4. Feng, N.S., Hahn, E.J.: Vibration analysis of statically indeterminate rotors with hydrodynamic bearings. J. Tribol.-Trans. ASME **120**(4), 781–788 (1998)
5. Wang, J.F., Sun, K.: Rotor critical speed calculation based on finite element method of a mixed model. Mach. Mater. Sci. Eng. Appl. **510**, 244–248 (2012)
6. Li, Z., Wang, Q.S., Qin, B., Shao, W.: Vibration characteristic analysis of flexible rotor-bearing system subjected to external combined loads. Eur. J. Mech. A-Solids **96**, 104688 (2022)
7. Gatilov, S.Y.: Using low-rank approximation of the Jacobian matrix in the Newton-Raphson method to solve certain singular equations. J. Comput. Appl. Math. **272**, 8–24 (2014)

Design and Testing of Modular Repeatable Folding Off-Rail Sail

Kang Gong[1], Licheng Liu[1](✉), Dongdong Li[1], Baiqing Tang[2], and Gang Wang[2]

[1] Beijing Satellite Manufacturing Factory Co. Ltd., Beijing, China
llc_whut@163.com

[2] Hebei Collaborative Innovation Center for Micro and Nano Satellites, Langfang, China

Abstract. In this paper, to overcome the deficiencies of the existing technology for the off-orbit problem of satellites, a space sail unfolding device is proposed, which realizes the functions of large-scale flexible sail stowage during the launching stage and sail unfolding in the orbit, and the device has the features of modularized combined use, repeatable folding and passive drive. This paper designs the layout and structure of the de-orbiting sail, clarifies the working form of the de-orbiting sail, analyzes the influence of the de-orbiting sail's unfolding synchronization, conducts modal analysis of the de-orbiting sail's storage and unfolding states, monitors the operation and action of the mechanism through the feedback data of the sensors and carries out the simulation and experimental analysis of the de-orbiting sail, so as to complete the de-orbiting task for the reliability of the target.

Keywords: Repeatable spreading · locking release mechanism · feedback data · off-track tasks

1 Introduction

In recent years, with the development of microelectronics, microelectromechanics and new materials, advanced manufacturing and other high-tech, small satellite technology has gained rapid development, and has gradually become one of the engines leading the development of space technology [1–3], and the number of small satellites has shown explosive growth in the last few years, which is a serious threat to the safety of spacecraft in orbit [4–6]. In order to minimize the threat of the number of satellites to on-orbit spacecraft, spacecraft should be removed for re-entry burnup within 25 years after the end of the mission of a low-orbiting spacecraft.

Currently, the main de-orbiting methods for satellites include propulsion technology [7], inflatable balloon technology [8], electrodynamic tethers [9], and de-orbit sails [10]. However, due to the limitations of the size and resources of microsatellites, it is difficult to apply some de-orbiting techniques [11–13] that are applicable to conventional satellites. Among them, the propulsion technology takes up some extra propellant on the star, which increases the mass of the satellite, and also requires continuous thrust vector control during the de-orbiting phase, which puts a higher demand on the lifetime

© The Author(s), under exclusive license to Springer Nature Singapore Pte Ltd. 2025
J. Mo and P. Chutima (Eds.): ICMSC 2024, LNME, pp. 41–52, 2025.
https://doi.org/10.1007/978-981-96-5380-5_5

of other subsystems on the star, especially the power supply and solar panels. Inflatable balloon technology requires a certain amount of pressurization device to be carried on board, which brings certain safety risks to the satellite launch and long-time work, and the unavoidable gas leakage caused by high-pressure gas also shortens the service life of this technology. The electric power tether is to use the high-speed movement of the conductive rope to cut the magnetic lines of the geomagnetic field to generate the force perpendicular to the direction of the tether and the local magnetic field, for some polar orbit satellites, this device has certain limitations.

The de-orbiting sail is mainly used to accelerate the de-orbiting of satellites by expanding the large-area sail surface in orbit to increase the resistance of satellites, and the de-orbiting sail can rely on its own stored mechanical energy to unfold, and the whole device is light in quality, simple in structure and low in cost, which is suitable for the rapid response and the micro-satellite with a short mission cycle.

2 Off-Rail Sails Work Process and Program Introduction

The de-orbiting sail is used to accomplish the de-orbiting task at the end of satellite operation in orbit, and its working form includes the ground stage, the launch stage, and the in-orbit stage, of which the in orbit stage includes the locking state in the early stage and the unfolding state at the end of satellite operation, as shown in Fig. 1.

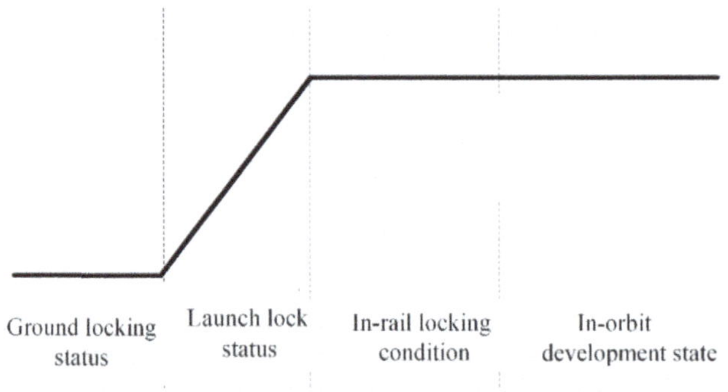

Fig. 1. Off-rail sails in different states

When the off-orbit sail is in the first three states, the elastic mast is wrapped around the center reel, the sail surface is stored in the sail box, and the sail cover is closed, During on-rail unfolding, the off-rail sail unfolds as shown in Fig. 2:

In the off-orbit sail installation position selection, the sail surface is configured at one end of the satellite, so that the center of mass of the satellite and the center of pressure of the sail surface offset to reach the maximum, the restoration torque generated by the satellite to reach the maximum value, the maximum to maintain the off-orbit sail's normal vector and the satellite's velocity vector is close to parallel to the sail surface in the generation of the maximum effective area of the air resistance, the principle of the principle is shown in Fig. 3:

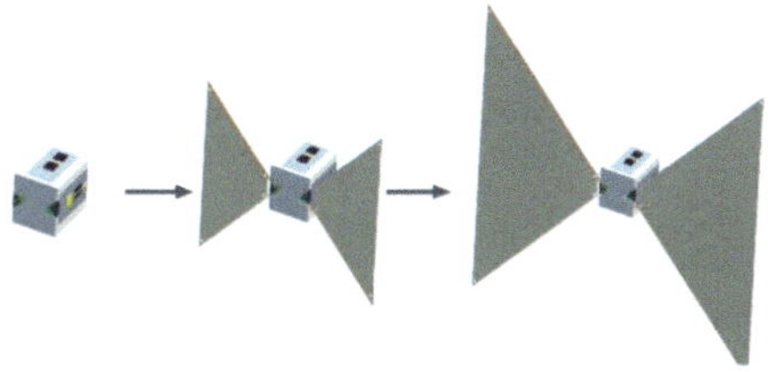

Fig. 2. Off-orbit sail deployment process

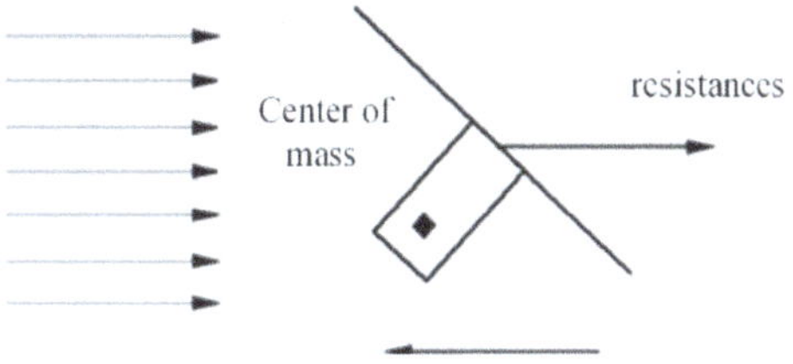

Fig. 3. Off-rail canvas position

In the layout design, the de-orbiting sail is placed as close as possible to the region of the satellite tail, so that according to the above principle, it can ensure that the de-orbiting sail plays the greatest role in the process of satellite de-orbiting.

3 Off-Rail Sail Structural Design

The modular off-rail sail is designed as shown in Fig. 7, and the whole device consists of an elastic mast assembly, a center reel, a guide wheel assembly, a sail surface box assembly, a sail cover and other components. Among them, the elastic mast provides power for the unfolding of the off-rail sail and maintains the sail surface in an unfolded shape after unfolding; the center reel is a storage device for the elastic mast; the guide wheel assembly mainly ensures that the elastic mast can be tightly wound around the center reel; the sail surface box assembly is mainly used for storing the sail surface; the guiding device is used to maintain the exhibiting direction of the mast and reduce friction; and the sail cover assembly is a locking device, which closes the elastic mast and the sail surface in the unit.

Combined with the main components of the off-rail sail module in Fig. 4 and the functional requirements of each component, the off-rail sail module is modeled and designed in detail, and the final off-rail sail overall shape three-dimensional model is shown in Fig. 5.

Including thickness t, angle θ, radius of curvature r, arc length b. These geometric elements will affect the mechanical properties of the strip spring, in order to determine

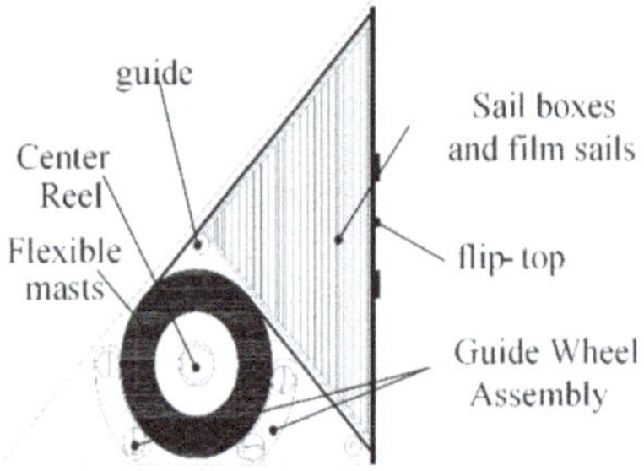

Fig. 4. Off-rail sail internal structure

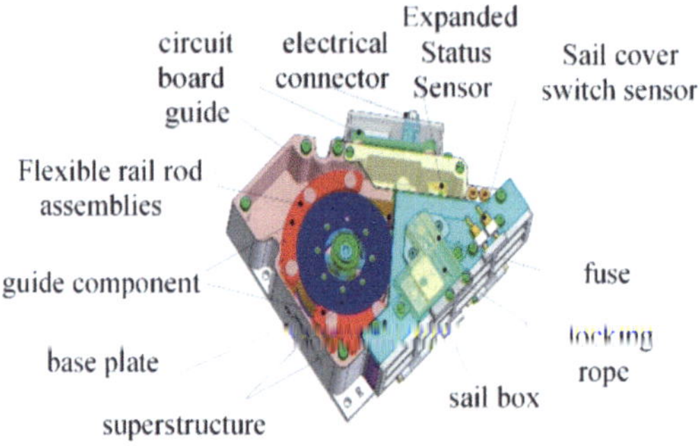

Fig. 5. 3D model of the off-orbit sail as a whole.

the cross-sectional parameters of the elastic mast, it is necessary to establish a mechanical model of the strip spring, the performance of the strip spring is the basis of the dynamics of the elastic mast analysis, due to the strip spring is a single layer of the open cylindrical shell structure, when the direction of the bending moment is different, their mechanical properties will also be There is a great difference in their mechanical properties.

The bending moment per unit arc length of a thin shell according to the thin shell bending theory of elastodynamics:

$$\begin{aligned} M_x &= D(\Delta k_x + \nu \Delta k_y) \\ M_y &= D(\Delta k_y + \nu \Delta k_x) \end{aligned} \tag{1}$$

where and are the curvature changes in the transverse and longitudinal directions of the strip spring, respectively; D is the bending stiffness of the thin shell, which can be obtained from the modulus of elasticity *E of* the thin shell spring, Poisson's ratio ν, and the thickness t of the thin shell:

$$D = \frac{Et^3}{12(1 - \nu^2)} \tag{2}$$

The curvature of the cross-section of the strip spring changes from $1/r$ to 0 when it is bent in the forward direction, while the curvature changes from 0 to $1/R$ along the length direction, as shown in Fig. 6.

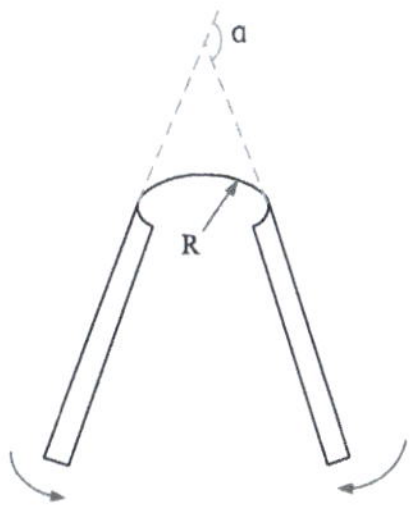

Fig. 6. Bending diagram of a strip spring

From the derivation, the strain energy in bending of the strip spring can be obtained:

$$\begin{aligned} U_{+} &= \frac{r\alpha\theta}{2} D\left[\frac{1}{R} + \frac{1}{r^2} + \frac{2\nu}{r}\right] \\ U_{-} &= \frac{r\alpha\theta}{2} D\left[\frac{1}{R} + \frac{R}{r^2} - \frac{2\nu}{r}\right] \end{aligned} \tag{3}$$

The unfolding moments of the strip spring for reverse and forward bending are, respectively:

$$\begin{aligned} M_{+}^{*} &= M_x r\theta = (1+\nu) D \frac{b}{r} \\ M_{-}^{*} &= M_x r\theta = (1-\nu) D \frac{b}{r} \end{aligned} \tag{4}$$

It can be seen that if the arc length b is a constant value, the magnitude of the unfolding bending moment of the ribbon spring is only related to its radius of curvature r. The influence of different section parameters on the strain energy of the ribbon spring can be obtained by the above equation, and the influence of carrying out each section parameter on the bending strain energy provides guidance for the selection of the design parameters of the mast section.

The whole assembly mainly consists of components such as center reel, elastic mast, pressure plate and bearing. The elastic mast is connected to the center reel through the pressure plate, and the center reel is connected to the top cover and the bottom plate through the bearings at both ends respectively.

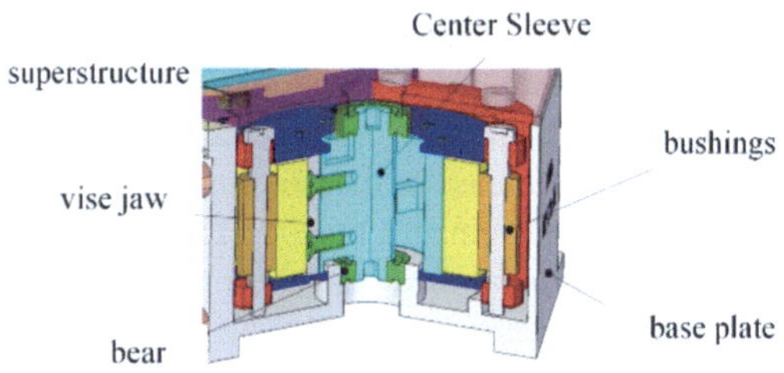

Fig. 7. Elastic mast winding assembly

The mast compression mechanism mainly consists of two sets of guide wheels, which are distributed along the circumference of the center reel, and respectively limit the outer ring of the elastic mast to ensure the elastic mast is normally curled and smoothly exhibited. The assembly mainly consists of guide bracket, guide spindle, guide slip sleeve and pad ring (Fig. 8).

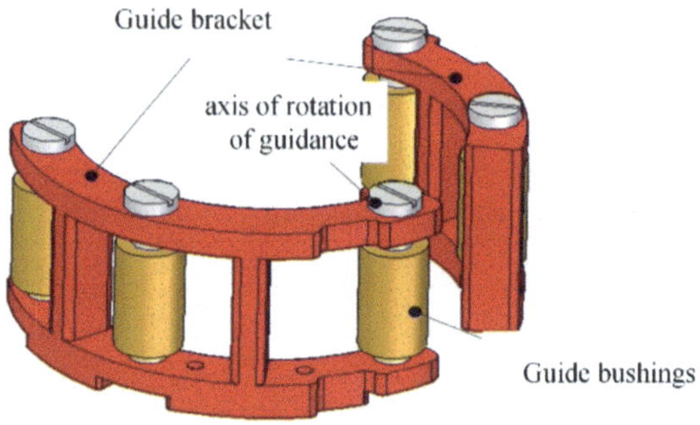

Fig. 8. Preload Wheel Assembly

In order to ensure the smooth extension of the mast and reduce the sliding friction in the extension section, a guide device is set up, which is mounted on the outer side of the sail box, and consists of a guide rod, bearing and sleeve, which mainly guides the elastic mast on one side (Fig. 9).

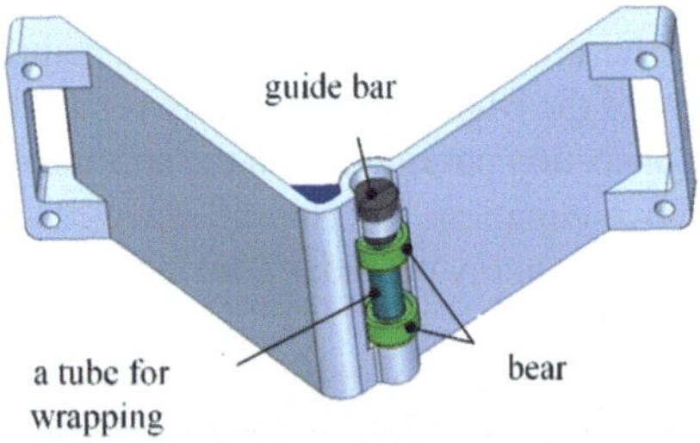

Fig. 9. Guides

The resistance wire is made of Cr20Ni80 NiCd wire, which has excellent resistance performance and is a mature heating material. The fuse rope is made of high strength polyarylate fiber. In the locking state, the two fuse ropes independently tighten the sail cover; when the power is released, the two ropes are heated at the same time, and both of them are fused to complete the release of the sail cover (Fig. 10).

For the satellite de-orbiting requirements, two de-orbiting sail modules are installed. Considering the stress concentration at the connection between the sail surface and the sail box, the shape of the sail surface is improved.

Off-rail sails have a large surface area and must be folded in the proper steps to reduce the size of the sail in the storage space. The choice of the proper folding method

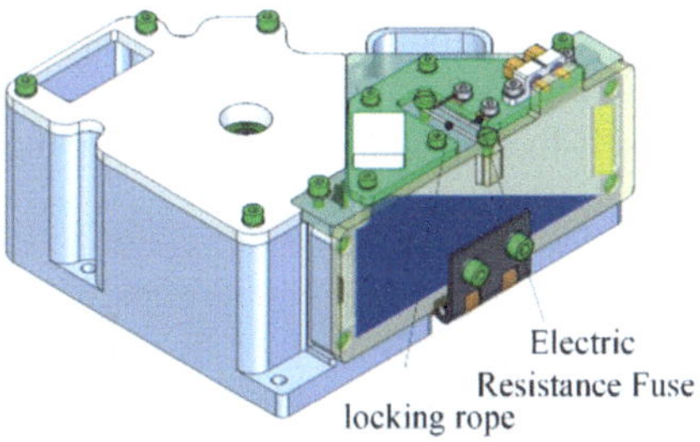

Fig. 10. Schematic diagram of electric resistance fuse

is very important and will determine whether the sail will "jam" or tear the film sail during unfolding. The way the sail is folded is determined by the way the sail is stored and the shape of the sail (Fig. 11).

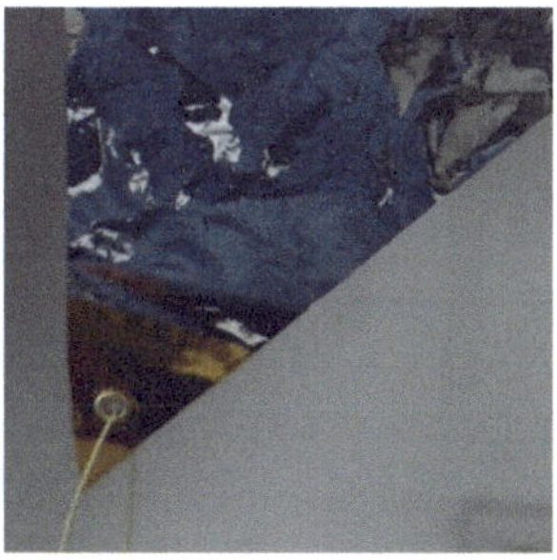

Fig. 11. Sail reinforcement test piece

In the storage state, the sail surface is folded and stored into the storage cavity as required and is flexibly connected to the elastic mast. The sail cover is locked by the locking rope to ensure that the shape of the sail remains unchanged and the flexible mast does not extend. When unfolding is required, the locking rope breaks and the sail cover opens by means of a torsion spring hinge.

4 Simulation and Experimental Analysis of Off-Orbit Sail Structures

To study the effect of different cross-section parameters on the strip spring, take the thickness t as an example to illustrate, when the thickness t is equal to 0.1 mm, the forward critical bending moment M of the strip spring is equal to 68.27 N·mm, and the reverse critical bending moment M is equal to 151.98 N·mm, when the thickness t is equal to 0.15 mm, the forward critical bending moment M of the strip spring is equal to 227.8 N·mm, and the reverse critical bending moment M is equal to 338.7 N·mm, and when the thickness t is equal to 0.2 mm, the forward critical bending moment M of the strip spring is equal to 492.77 N·mm, and the reverse critical bending moment M is equal to 594.47 N·mm.

As the thickness increases, the critical bending moment increases. When it is necessary to increase the ability of the strip spring to resist interference after self-locking, it is possible to increase its self-locking ability by increasing the thickness (Figs. 12, 13 and 14).

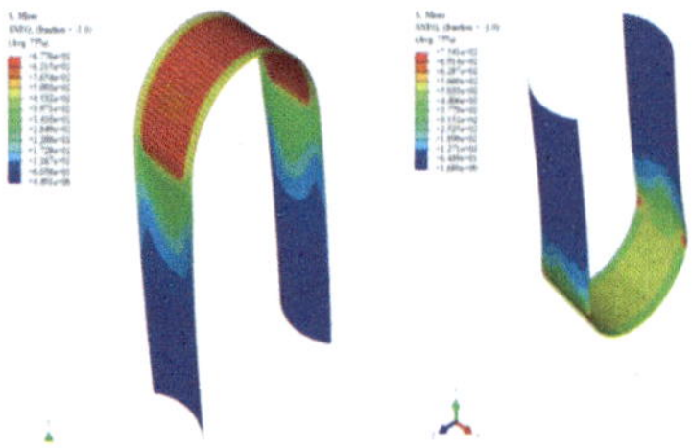

Fig. 12. t = 0.1 mm, bending strain diagram of reed

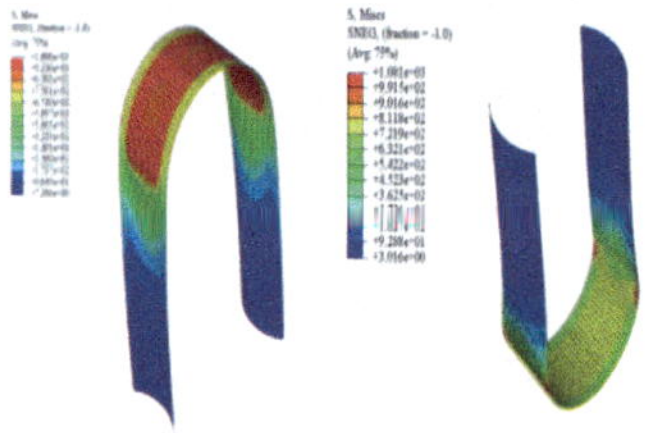

Fig. 13. t = 0.15 mm, bending strain diagram of reed

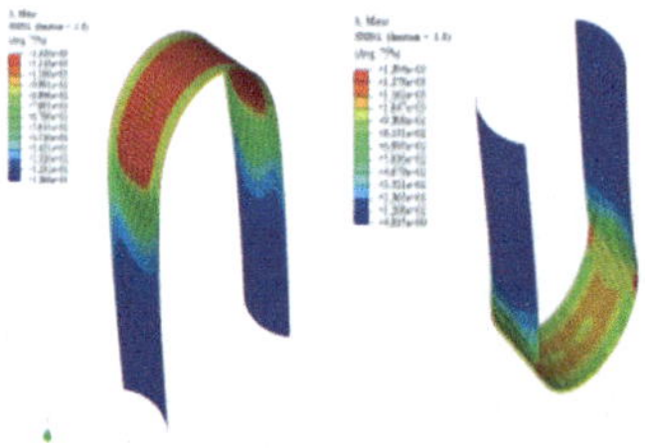

Fig. 14. t = 0.2 mm, bending strain diagram of reed

Thickness on the strain energy, when the thickness t is equal to 0.1 mm, the band spring forward bending strain energy is equal to 62.85 mJ, reverse bending strain energy reverse critical moment is equal to 130.54 mJ, when the thickness t is equal to 0.15 mm, the band spring forward bending strain energy is equal to 212.12 mJ, reverse bending strain energy reverse critical moment is equal to 440.57 mJ, when the thickness t is equal to 0.2 mm, the band spring forward bending strain energy is equal to 502.81 mJ, reverse bending strain energy reverse critical moment is equal to 1044.3 mJ. When the thickness t is equal to 0.2 mm, the forward bending strain energy of the strip spring is equal to 502.81 mJ, and the reverse bending strain energy of the strip spring is equal to the critical bending moment in the reverse direction is equal to 1044.3 mJ.

With the increase of thickness, there is a large increase in the strain energy of the strip spring, which shows that thickness is a sensitive factor affecting the strain energy. Increasing the thickness can effectively increase the strain energy and improve its automatic unfolding ability.

Effect of thickness on unfolding bending moment:

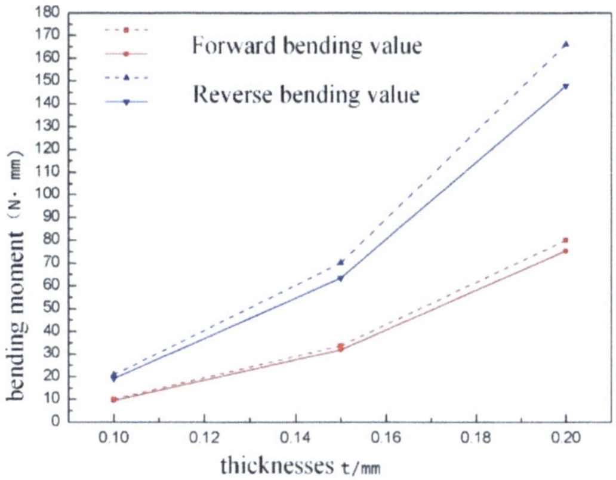

Fig. 15. Effect of thickness t on unfolding moment

As can be seen from Fig. 15, its unfolding bending moment numerical simulation analysis value is not much different from the theoretical calculation value, and the calculation result is better matched with the theoretical value, and the unfolding bending moment of the strip spring has a large increase with the increase of thickness. Increasing the thickness can effectively increase the unfolding moment and improve its automatic unfolding ability. In the same way, numerical simulations are carried out for the effects of changes in wrap angle θ, radius of curvature r, arc length b, which are in good agreement with the theoretical calculations and are not repeated here.determined by the mast, the mast modes were analyzed, and the modal results are shown in Fig. 16.

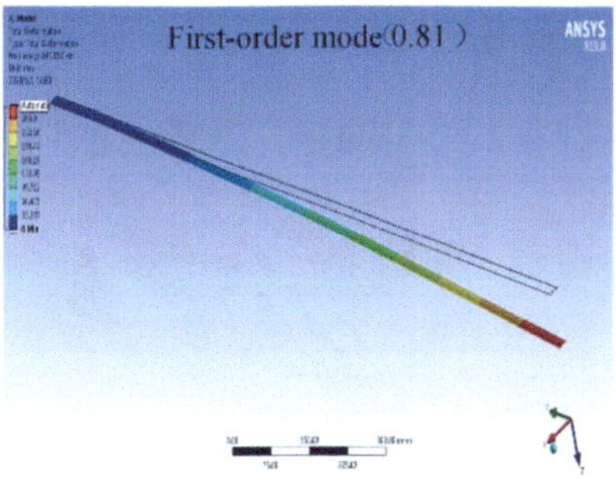

Fig. 16. Mast 1st order fundamental frequency

The elastic mast was subjected to a thrust test at the front end of the mast before surface coating with molybdenum disulfide and when assembled to the official product, as shown in Fig. 17, to obtain the magnitude of the thrust at the outstretched end of the mast as it progresses from stowed to unfolded.

In order to compare the effect of sail reinforcement, firstly, the test sample without hollow rivet protection was subjected to tensile force test, when the test tensile force was 5N, small deformation occurred at the location of the connection port, when the test tensile force was 10N, obvious deformation occurred at the location of the connection port, and when the test tensile force was 15N, the location of the connection port was instantly fractured.

The film reinforcement area range is slightly different, the test test tension direction is different, making the film maximum bearing capacity is different, the maximum bearing capacity is greater than no hollow rivets.

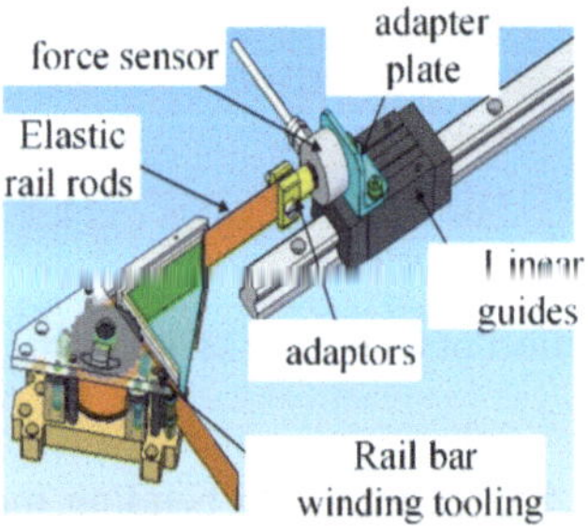

Fig. 17. Mast front thrust test

Unlike the stiffness of paper, double-sided aluminized polyimide film has a certain degree of elasticity, so only folded without pressure, the crease can not be left, therefore, the design of the folding/pressure of the sail surface of the tooling, as shown in Fig. 18, the tooling has a certain degree of expandability, after the folding is completed, the use of vacuum pressurization.

Fig. 18. Membrane after first and second folding

Resistance wire fusing device in the fusing rope using polyarylate fiber Vectran, its role is to lock the sail cover, in the rail locking stage has been in the state of bearing tension, the design of the fiber strength is fully considered.

5 Conclusion

In this paper, a space sail unfolding device is proposed, firstly, the working process of off-orbit sail is introduced, then the structure of off-orbit sail is designed, which realizes the function of large-scale flexible sail stowage during launching stage and sail unfolding in orbit, and the device has the characteristics of modular combination, repeatable folding and passive drive, the structural strength of the off-orbit sail is analyzed by simulation, and finally the rail bar of the off-orbit sail is tested by thrust test, which meets the target requirements. Finally, the thrust test of the de-orbiting sail is carried out to meet the target requirements. The designed de-orbiting sail can be repeatedly unfolded and stored on the ground, and the sail surface can be reliably unfolded under the action of the unfolding mechanism, and the fixed unfolding state can be maintained in space until the satellite completes the de-orbiting task, so that it can fulfill the de-orbiting task reliably for the target.

Acknowledgments. Kang Gong would like to thank the National Defense Innovation Fund Project of North China Institute of Aerospace Industry (project no. GFCXJJ-2024-02), the Funded by Science Research Project of Hebei Education Department (project no. ZD2022113), the Central Guidance on Local Science and Technology Development Fund of Hebei Province (project no. 226Z1803G). Acknowledges the financial support from The Natural Science Foundation of China (Projects No. 52305002).

References

1. Zhang, Q., Zhang, G., Ding, X.: Special issue on tunnel construction and underground space technology. Buildings **13**(12), 2988 (2023)
2. Space Technology Graduate Research Opportunities–Fall 2024 (NASA). Federal Grants & Contracts **47**(20), 6 (2023)
3. GKN Aerospace unveils Global Technology Centre (GTC) in the Netherlands, advancing sustainable aerospace technology. M2 Presswire (2023)
4. Ma, X., Hu, Y., Yu, M., et al.: Orbit determination of high orbit spacecraft based on BDS-3 using extended Kalman filter. J. Phys. Conf. Ser. **2762**(1) (2024)
5. Fei, J., Li, X., Zhang, Z.: Research on teleoperation service efficient scheduling technique of long-term on-orbit spacecraft. Acad. J. Comput. Inf. Sci. **2**(3) (2019)
6. Patents; Researchers Submit Patent Application: Inclined Geosynchronous Orbit Spacecraft Constellations, for Approval (USPTO 20170247123). Defense & Aerospace Week, 316-(2017)
7. Mechanical Engineering; Investigators at Korea Advanced Institute of Science and Technology (KAIST) Detail Findings in Mechanical Engineering (Assessment of architectural options for a dual-mode disaster monitoring constellation supported by on-orbit propellant depots). J. Eng. (2014)

8. Koryanov, V.V., Toporkov, A.G., Kazakovtsev, V.P., Nedogarok, A.A.: The technology of using inflatable structures made of special materials to adapt the de-orbiting system for removing spacecraft. IOP Conf. Ser. Mater. Sci. Eng. **812**(1), 012006 (2020). https://doi.org/10.1088/1757-899X/812/1/012006
9. Valmorbida, A., Olivieri, L., Brunello, A., Sarego, G., Sánchez-Arriaga, G., Lorenzini, E.C.: Validation of enabling technologies for deorbiting devices based on electrodynamic tethers. Acta Astronautica **198**, 707–719 (2022)
10. Yulei, F., Ruifeng, G., Zhiang, D., et al.: Study of the deorbit sail damage under the combined effects of atomic oxygen erosion and material defects. Acta Astronautica **213**, 741–751 (2023)
11. Sasaki, T., Nakamura, R., Okamoto, H.: Optimal attitude planning along torque equilibrium points using aerodynamic database for deorbiting large space debris. Acta Astronautica **220**, 243–253 (2024)
12. Palla, C., Kingston, J.: Forecast analysis on satellites that need de-orbit technologies: future scenarios for passive de-orbit devices. CEAS Space J. **8**(3), 191–200 (2016)
13. Kingston, J., Hobbs, S., Roberts, P., et al.: Use of CYPRES™ cutters with a Kevlar clamp band for hold-down and release of the Icarus De-Orbit Sail payload on TechDemoSat-1. Acta Astronautica **100**, 82–93 (2014)

Design and Test of Heavy-Duty Low-Impact Unlocking Device Based on Line Constraints

Licheng Liu[1], Kang Gong[1(✉)], Yanni Xiao[1], Baiqing Tang[2], and Gang Wang[2]

[1] Beijing Satellite Manufacturing Factory Co., Ltd, Beijing, China
Gongkang86@163.com

[2] Hebei Collaborative Innovation Center for Micro and Nano Satellites, Langfang, China

Abstract. This paper puts forward the separation and unlocking technology of line constraint for the problem of large separation impact and weak testability of traditional pyrotechnic separation device, which can not only ensure the reliable connection and separation of components, but also ensure that the impact of separation is as small as possible, reduce the impact of separation impact on the equipment load, and satisfy the complexity of the aircraft and the large load demand, by analyzing the factors affecting the winding force of the spring of the line constraint, and guiding the design of the device on By analyzing the factors affecting the winding force of the wire constrained spring, the design of the device is guided to control these factors, so that the load carrying capacity, separation impact, and other performance indicators of the separation device can meet the requirements of large load carrying, low impact, and fast separation technology. The trigger mode has an important impact on the power consumption and unlocking time of the device, and the trigger mode is selected through mechanical analysis and layout design, so that the separation device can meet the requirements of low-power consumption and fast response. The principle prototype is designed, and the feasibility and performance of the principle prototype are analyzed through simulation and experimental testing to provide comprehensive technical support for its space engineering applications.

Keywords: Connection and separation · triggering mode · mechanical analysis · principle prototype design

1 Introduction

Connection and separation technology [1–3] is a key product to realize satellite launch and separation in orbit, and is widely used in the fields of inter-stage separation of satellites and rockets, segment separation, star-arrow separation and load release. Connection and separation device realized in the vehicle launch and flight process, will need to satellite or load and flight platform reliable lock into one, when receiving the separation command, to ensure that the two parts of the structure can be connected to the safe and accurate separation. Therefore, the connection and separation device and the connection and separation technology it belongs to occupy an important position in the system engineering of air and space vehicles, and it is one of the key technologies to ensure the success of the mission.

© The Author(s), under exclusive license to Springer Nature Singapore Pte Ltd. 2025
J. Mo and P. Chutima (Eds.): ICMSC 2024, LNME, pp. 53–64, 2025.
https://doi.org/10.1007/978-981-96-5380-5_6

At present, most of the satellite connection separation devices applied in the field of aerospace [4–6] are based on the principle of pyrotechnic action, i.e., cutting, punching off or pushing out the connecting parts through the explosion impact force to realize separation. The pyrotechnic separation [7] device has the advantages of simple structure, strong bearing capacity, rapid separation action, etc. It has been tested by a large number of flight tests and is a mature stand-alone product. However, such devices also have insurmountable shortcomings, mainly including large additional weight, strong separation impact, obvious additional pollution, non-reusable, etc., which seriously limits their application in high-precision, low-impact separation tasks. With the improvement of separation control precision of on-orbit cluster launch of future microsatellites [8–10] as well as the wide application of various types of sensitive equipments (e.g., optical cameras, thin-film mirrors, flexible antennas, etc.) with harsh requirements on the impact environment, the limitations of the existing pyrotechnic separation device technology will be more and more obvious. Therefore, for the future of one-arrow multi-satellite in-orbit launch mission [11] to promote microsatellite technology as soon as possible to form the equipment capacity, there is an urgent need to develop a new type of standardized, low-impact connection and separation device, to cope with the increasing demand for 1–100 kg microsatellite in-orbit intensive launch, and to promote its space engineering applications.

The study of the line-constrained heavy-duty low-impact resetable unlocking technology based on elastic elements will be used as a new type of separation and unlocking method. A new type of separation unlocking method to provide technical support for the future development of space technology.

2 Heavy Duty Low Impact Unlocking Technology Program Design

The unlocker, shown in Fig. 1, is mainly composed of fused wire unlocking mechanism, compression top plate, compression bottom plate, preload lever, lock nut, separation spring, capture cap and other parts.

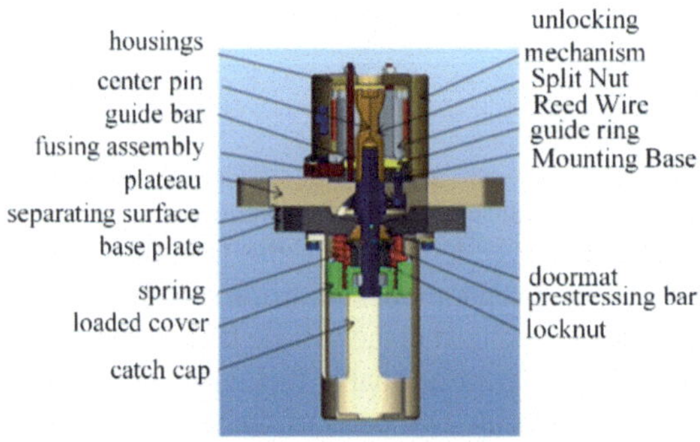

Fig. 1. Unlocking mechanism

Locking state: the fused wire unlocking mechanism is installed in the compression upper plate, its internal center insertion rod and preload rod as the main force device, the upper end of the preload rod and the center insertion rod through the threaded connection, the lower end of the use of the lock nut for fastening, the center insertion rod external by the flap slider for fixing, at the same time, the use of constraints spring wire winding way to hold the flap slider tightly, constraints reed wire is fixed on the top end of the flap slider at one end, and the other end is pulled tight by the fused wire of the fuse assembly. One end of the restraining reed wire is fixed to the upper end of the flap slider, and the other end is tightened by the fuse wire of the fuse assembly. After applying the preload force, the device locks the top plate and bottom plate to complete the locking of the separation surface.

The fused wire unlocking mechanism is an important part of the whole unlocker to realize the carrying and unlocking functions. Analyzing the functions and performance indexes of the unlocker, the detailed design of the fused wire unlocking mechanism is carried out, and the composition of the fused wire unlocking mechanism is shown in Fig. 2:

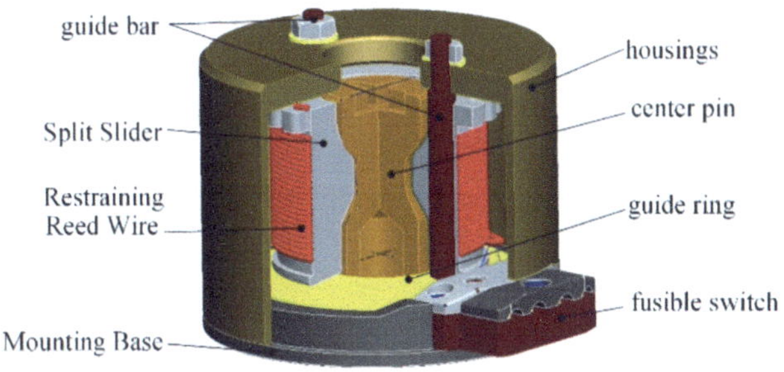

Fig. 2. Schematic diagram of fused wire unlocking mechanism

The carrying part of the fused wire unlocking mechanism is mainly composed of flap slider, center rod, constraint reed wire, the main function is to transform the axial bearing force into a relatively small radial constraint force, triggering the unlocking function is completed by the fusible switch to release the constraint force of the constraint reed wire by the fusible fuse; in addition, the unlocking device also includes the mounting base, shell, guide bar, and other parts, which are mainly for the unlocking device to provide the mounting as well as protection Function.

The fused wire unlocking mechanism from the locking state to the unlocking state mainly consists of two steps, first of all, the fused wire is energized and heated up to melt, and the restraining reed wire expands outward under the action of its own restoring force; and then, the flap slider separates under the action of the outward force of the central insertion rod, lifting the restriction on the central insertion rod, and the central insertion rod completes the release.

3 Mechanical Analysis of Wire-Constrained Spring Windings

Line constrained spring winding bearing technology, mainly using the radial constraint force generated in the process of constrained spring wire winding to limit the movement of the split flap nut, to achieve the role of bearing.

According to the static equilibrium, it is obtained:

$$\begin{aligned} dN &= 2T\sin\frac{d\alpha}{2} + dT\sin\frac{d\alpha}{2} \\ (T + dT)\cos\frac{d\alpha}{2} &= T\cos\frac{d\alpha}{2} + \mu dN \end{aligned} \tag{1}$$

Since *da* converge to 0, the above equation simplifies to give:

$$\begin{aligned} dN &= Td\alpha \\ dT &= \mu dN \\ T_1 &= Te^{\mu\alpha}, T = T_1e^{-\mu\alpha} \end{aligned} \tag{2}$$

Therefore, the friction force and the circumferential positive pressure can be obtained:

$$\begin{aligned} f &= T_1 - T = T_1(1 - e^{-\mu\alpha}) \\ P &= {}^{f}/_{\mu} \end{aligned} \tag{3}$$

The relationship between the force T at the free end of the restraining reed wire and the radial force P of the flap slider can be derived from the above equation, as shown in the following equation:

$$\frac{T}{P} = \frac{\mu}{e^{\mu\alpha} - 1} \tag{4}$$

Through the formula, it is found that the force on the free end of the restraining spring wire depends on the number of winding turns of the restraining spring wire and the friction coefficient between the restraining spring wire and the split-valve slider. Therefore, the design of the number of winding turns of the constrained spring wire and the friction coefficient between the constrained spring wire and the split-valve slider can be used to achieve the effect of force increase, i.e., when the constrained spring wire is wrapped around the split-valve slider, the split-valve slider can withstand a larger load carrying capacity by means of the smaller constrained wire constraining force.

The main characteristic of a fuse wire is that when the current is too high, the heat generated by the fuse wire due to its own resistance is very high, and when the temperature exceeds the melting temperature of the fuse wire, the fuse wire will break. By utilizing this characteristic of the fuse wire, it can be designed as a trigger drive element for the unlocking device. In the case of a load-bearing structure using a wire-constrained spring winding, the free end of the constrained spring wire can be restricted by the fusing wire during the locking state, and the unlocking function can be realized by contacting the restriction of the constrained spring wire when the fusing wire is energized to fuse.

Fig. 3. Stainless steel wire pull-off force test

Stainless steel wire was selected as the material of the fusing wire, and the pull-off force of stainless steel wire of different diameters was measured (Fig. 3):

In order to test the performance of stainless steel wire energized fusion, selected 0.1 mm, 0.12 mm, 0.15 mm, 0.2 mm four different specifications of stainless steel wire for testing, the length of the steel wire is 80 mm, energized current of 2 A and 4 A test is shown in the figure below (Fig. 4):

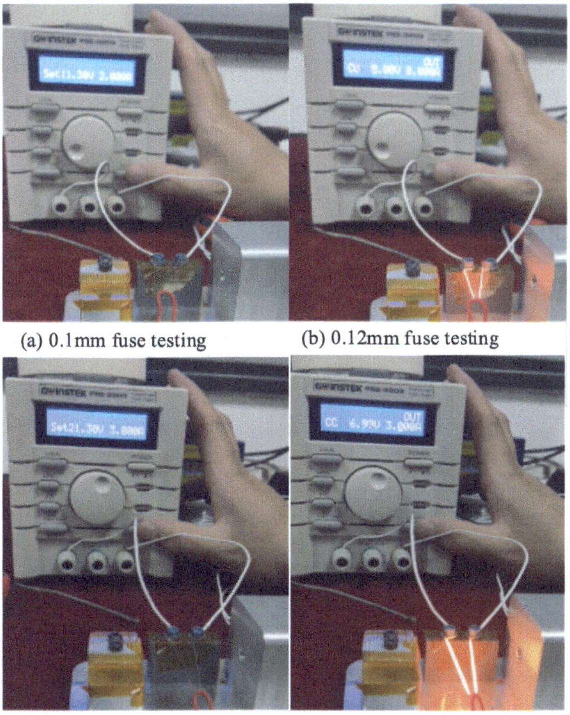

(a) 0.1mm fuse testing (b) 0.12mm fuse testing

(c) 0.15 fuse wire test 2A current (d) 0.15 fuse wire test 3A current

Fig. 4. Stainless steel wire fusion test

Test results show that the diameter of 0.1 mm, 0.12 mm in 2 A current can be instantaneous fusion, diameter 0.15 mm in 3 A current can be instantaneous fusion, diameter 0.2 mm fuse wire in 2 A, 3 A, 4 A current can not be quickly fused. Therefore, suitable for use as a trigger drive stainless steel wire diameter should be less than 0.15 mm.

Through the experimental results, it is known that the fuse wire has the characteristics of small size, fast trigger time, light weight and strong environmental adaptability. Therefore, in order to adapt to the requirements of low impact and fast response in the wide temperature range of future spacecraft, the fuse wire has obvious advantages.

4 Schematic Prototype Design and Performance Testing

4.1 Chematic Prototyping

Unlocker from the locking state to the unlocking state mainly contains two steps, first of all, the fuse wire energized sound heating melting, constraint spring wire in its own response force under the action of outward expansion; then, the split flap nut in the center of the center of the rod under the action of the outward force separation, contact on the center of the rod limitations of the center of the rod, the center of the rod to complete the release of the release of the unlocker bearing part is mainly composed of the split flap slider, the center of the rod, the constraint spring wire. Constraint spring wire for the flap slider to provide radial binding force, so that the flap slider to limit the axial movement of the center rod, so as to achieve the bearing, through the design of the flap slider and the center rod to match the bearing structure. Carrying, through the line constraint flexible spring (referred to as the constraint reed wire) so that the split-flap slider as a whole, to limit the release of the center rod; unlocked, the constraint reed wire release, lifting the restriction on the split-flap slider, the split-flap slider apart, so that the center rod out of the center to achieve the unlocking of the separation, as shown in the figure below (Fig. 5):

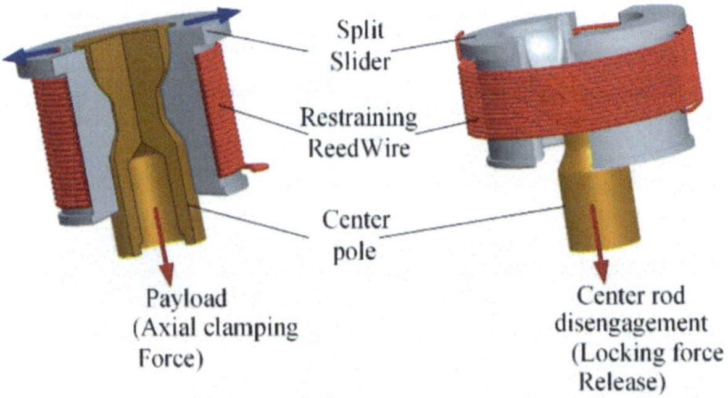

Fig. 5. Line Constraint Separation Unlocking Schematic

The center rod is connected with the preload rod through threads, and the loading nut is also connected with the preload rod through threads. Therefore, the thread of the preload rod to carry out the selection of design, connecting screw bearing capacity calculation formula:

$$F = \pi (D/2)^2 \sigma \tag{5}$$

The minimum bolt diameter can be calculated as:

$$D \geq 2\sqrt{F/_{\pi\sigma}} \tag{6}$$

Check the bolt standard can be seen, M4 × 0.7 thread small diameter of Φ3.242, so can be connected to the bolt for M4 × 0.7, check the calculation can be seen.

$$F_{M4} = \pi (D/2)^2 \sigma = 9.08KN \tag{7}$$

Calculate the axial force:

$$\begin{aligned} F &= N_0 \cdot \sin\theta + f_0 \cdot \cos\theta \\ N_0 &= \frac{F}{\sin\theta + \mu \cdot \cos\theta} \end{aligned} \tag{8}$$

It is possible to find the radial constraint force of the constraining reed wire limiting the flap slider

$$\begin{aligned} F_h + f_0 \sin\theta + f_1 \cos\beta &= N_0 \cos\theta + N_1 \sin\beta \\ N_0 \sin\theta + f_0 \cos\theta &= N_1 \cos\beta + f_1 \sin\beta \end{aligned} \tag{9}$$

Contact area and Contact Stress:

$$\begin{aligned} S &= l \cdot c \\ \sigma &= \frac{N_0}{S} \end{aligned} \tag{10}$$

The restraining spring wire involves the following conditions to be mainly satisfied: ensure that the unlocking state does not affect the opening of the flap slider, the locking state can reliably hoop the flap slider to unfold, and that the restraining spring wire does not yield in the hooped state.

The unlocker mainly includes the bearing part, the trigger unlocking part, and the mounting shell part, which can be divided into 3 major components, and the bearing structure part is shown in the figure (Fig. 6):

Load-bearing structure, the center rod with internal threads, for the preloading of the compression structure, divided flap slider and the center of the rod between the arc contact, play a role in limiting, but also allows a certain degree of installation error; constraints on a section of the reed wire and the fuse wire connected to the other end of the hook in the mounting holes in the divided flap slider, to limit the radial displacement of the divided flap slider.

The structure of the unlocking part is shown in the figure (Fig. 7):

In the unlocking structure, the switch mounting seat insulation is used to insulate the fuse wire from the body of the unlocker, and the fuse wire is connected to one end of the restraining spring to limit the displacement of the restraining spring wire.

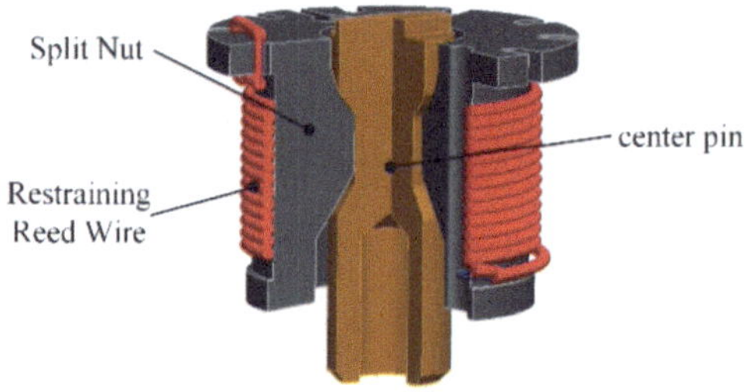

Fig. 6. Load-bearing structure

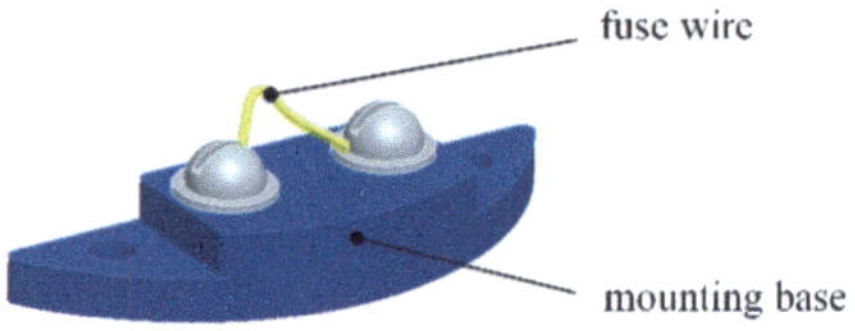

Fig. 7. Unlocking Structure

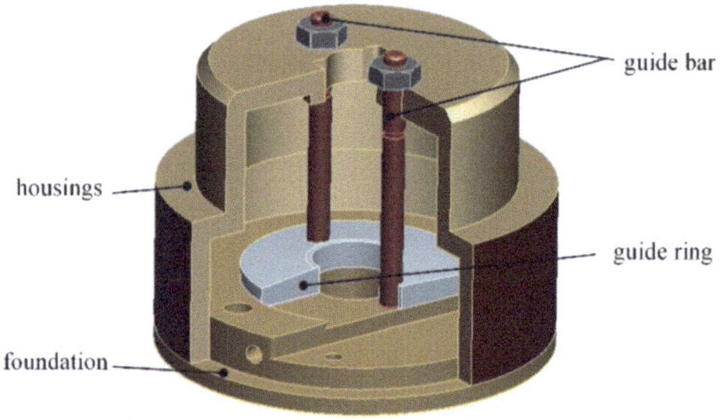

Fig. 8. Mounting housing structure

The mounting housing structure is shown below (Fig. 8):

The mounting shell structure is used to mount other components and provide support and protection, the bottom guide ring is used to guide the flap slider at the time of separation, and the guide bar can limit the rotation of the flap slider, and at the same time, it also has a positioning effect on the flap slider.

4.2 Simulation and Test Verification of Unlocking Device

After completing the design and theoretical analysis of the unlocker scheme, the static load capacity characteristics of the device are analyzed using the finite element analysis software Ansys, and the function and performance of the unlocking device are tested using experimental tests.

Due to the need to ensure that the unlocking device work when the load bearing capacity requirements, flap slider and the center of the rod through the arc contact, by the force transfer path can be seen in the structure of the largest load-bearing components for the flap slider and the center of the rod, therefore, the two parts for the calibration.

Through three-dimensional modeling and model simplification to remove unnecessary grooves and chamfers, simulation analysis, the use of solid units and the definition of material constants, the use of free mesh division method for solid division, the establishment of finite element model. Apply full constraints on the outer cylindrical surface of the flap slider, set up a friction pair of contact between the flap slider and the center rod, and apply an axial 6 kN preload on the center rod, and the results of its finite element analysis are shown in Fig. 9. From Figure (a) can see the maximum stress suffered by the flap slider is 756 MPa, the material of the flap slider is 50Cr, the tensile strength of 1100 MPa, the yield limit of 930 MPa, so there is enough strength margin; from Figure (b) can see the maximum stress of the center rod is 1087.2 MPa, the center rod material is TM210 A (lithium-aluminum alloy), and the yield strength is 1960 MPa, so the center rod has the strength to meet the requirements.

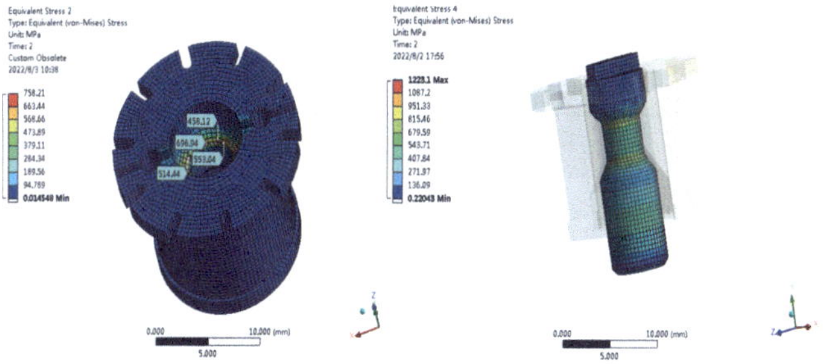

(a) flap slider (b) center inserted rod

Fig. 9. Mounting housing construction

In order to verify the unlocking impact of the line constrained bearing technology, the unlocking impact of the principle prototype was tested. The impact performance test method refers to the impact test of pyrotechnic separation, and the impact test platform is designed to be suspended and installed, with different distance measurement points in the center field and far field for impact data acquisition (Fig. 10).

Impact test equipment using unidirectional acceleration sensor (piezoelectric) frequency range of 1–5000 Hz, the magnitude of ± 1000 g, lateral sensitivity is better than 5%, the use of three-way acceleration sensor (piezoelectric) frequency range of 1–5000 Hz, the magnitude of ± 500 g, the transverse sensitivity is better than 5%, the use of modal test system VR3-100447 number of acquisition of 8 channels. VR3-100446 digital acquisition for 16 channels, the accuracy of 0.5% of the amplitude.

This unlocking impact test was carried out two tests, the sampling time of the impact test is 3 s, the state of the two test specimens and the test software parameter settings are the same, the test unlocking is shown in the figure below (Fig. 11).

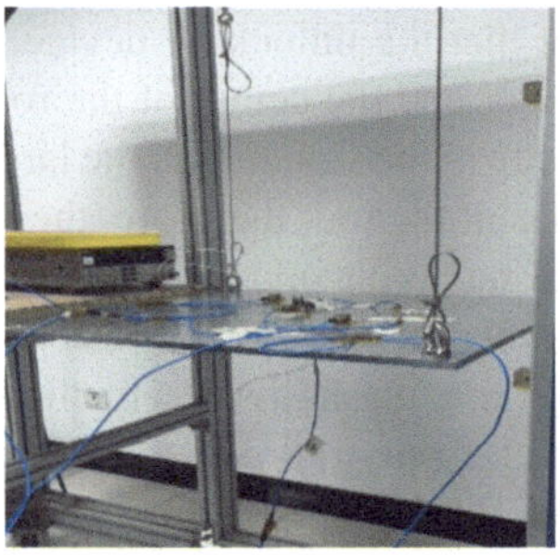

Fig. 10. Unlocking device impact test test

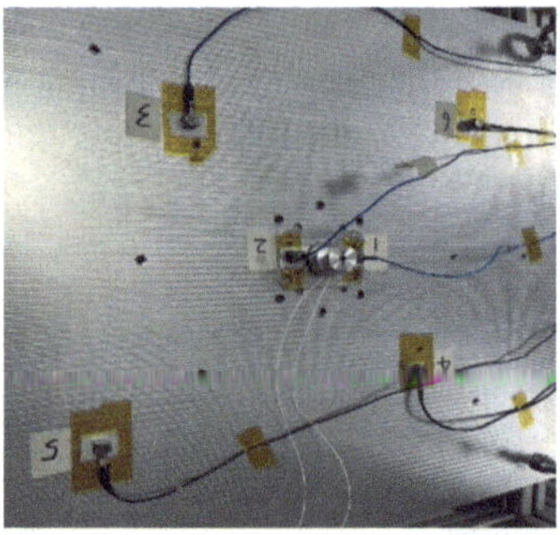

Fig. 11. Sensor Positioning Diagram

The first impact test, all the test points impact response time domain curve shown in Figure, the maximum response point is the No. 2 test point, the maximum value of impact acceleration is 256.96 g, the impact level is much smaller than the same level of fire unlocker.

In the second impact test, the impact response time domain curves of all measurement points are shown in Fig. The maximum response point is measurement point No. 2, and the maximum value of impact acceleration is 303.12 g, which is not much different from the first impact response test result, indicating that the performance of this unlocking device is relatively stable.

The unlocking shock magnitude of the fused-wire unlocking device based on the wire constraint winding is much smaller than that of the same magnitude pyrotechnic unlocker, which is more suitable for the unlocking and separating occasions of the future spacecraft that require low shocks.

5 Conclusion

The separation and unlocking technology with line constraints proposed in this project mainly focuses on the problems of large separation impact and weak testability of traditional pyrotechnic separation devices, which can ensure the reliable connection and separation of components, but also ensure that the impact during separation is as small as possible, reduce the impact of the separation impact on the equipment load, and satisfy the complexity of the vehicle and the large loading requirements. Analysis of the winding force of the wire constrained spring.By analyzing the factors affecting the winding force of the wire constrained spring, the design of the device is guided to control these factors, so that the load carrying capacity of the separation device, the amount of separation impact, and other performance indicators to meet the requirements of large load carrying, low impact, rapid separation technology; Trigger mode mainly affects the power consumption and unlocking time of the device, and at present, the trigger modes of the separation device mainly include shape memory alloy drive, electromagnetic drive, thermal drive, paraffin drive, motor drive, etc. Through mechanical analysis and layout design, the trigger mode is selected so that the separation device can satisfy the requirements of low-power consumption and fast response. Design the principle prototype, analyze the feasibility and performance of the principle prototype through simulation and experimental testing, summarize and accumulate design experience and theory, and provide comprehensive technical support for its space engineering applications.

Acknowledgments. Licheng. Liu would like to thank the National Defense Innovation Fund Project of North China Institute of Aerospace Industry (project no. GFCXJJ-2024-02), the Funded by Science Research Project of Hebei Education Department (project no. ZD2022113), the Central Guidance on Local Science and Technology Development Fund of Hebei Province (project no. 226Z1803G). Acknowledges the financial support from The Natural Science Foundation of China (Projects No. 52305002).

References

1. Castaneda, CA., Dao, TP., Yang, Y., Galagedera, S.K.K.: Mechanistic insights into the promotion or inhibition of phase separation by polyubiquitin chains of different linkages. Biophys. J. **121**(3), 146a (2022). https://doi.org/10.1016/j.bpj.2021.11.1989
2. Shenzhen Luxshare Precision Industry Co., Ltd.: Agency Reviews Patent Application Approval Request for Electrical Connector with Grounding Contact Having Forked Soldering Branches. Politics & Government Week (2013)
3. Gang, W., Yimeng, Y., Jingtian, W., et al.: Design of point-type heavy load connection and low impact separation mechanism based on energy flow analysis. Adv. Mech. Eng. **14**(12) (2022)
4. Careri, F., Khan, R.H.U., Todd, C., Attallah, M.M.: Additive manufacturing of heat exchangers in aerospace applications: a review. Appl. Therm. Eng. **235**, 121387 (2023). https://doi.org/10.1016/j.applthermaleng.2023.121387
5. Innodisk Introduces the First PCIe nanoSSD 4TE3 with Compact Size, Reliability and Performance to Unlock 5G, Automotive and Aerospace Applications. M2 Presswire (2023)
6. Joo, Y., Kim, C., Kang, Y., et al.: Carbon nanotube mat/boron nitride composite materials for aerospace applications. Surf. Coat. Technol. **479**, 130577 (2024)
7. Jeongmin, W., Ho, J.K., Yeon, J.C., et al.: Influence of design parameters on the behavior of pyrotechnic separation nut. J. Korean Soc. Aeronaut. Space Sci. **47**(9), 617–628 (2019)
8. Yin, C., Meng, T.: High frequency attitude determination of micro satellites based on FPGA. J. Phys. Conf. Ser. **2746**(1), 012042 (2024)
9. Mehdi, M.S.N.H., Jafar, R.: Innovative exploration of a bio-inspired sensor fusion algorithm: enhancing micro satellite functionality through Touretsky's decentralized neural networks. J. Intell. Robot. Syst. **110**(2) (2024)

10. Rigatos, G., Abbaszadeh, M., Busawon, K., et al.: Flatness-based control in successive loops for unmanned aerial vehicles and micro-satellites. Guid. Navig. Control **03**(04) (2024)
11. MUOS-3 satellite on track for launch despite hardware deficiency. Inside the Pentagon **30**(35), 13–13 (2014)

Development of Real-Time Leg Joint Angle Estimation System Using Single IMU and Deep Learning

Koyo Toyoshima, Jae Hoon Lee(✉), and Shingo Okamoto

Graduate School of Science and Engineering, Ehime University, 3 Bunkyocho, Matsuyama, Ehime, Japan
lee.jaehoon.mc@ehime-u.ac.jp

Abstract. The prevailing methodologies for gait analysis methods require expensive and large-scale motion capture equipment, which has led to increased interest in more accessible approaches using a small number of wearable sensors. This study proposes a method for real-time estimation of hip, knee, and ankle joint angles of both legs during walking using a single, relatively inexpensive Inertial Measurement Unit (IMU) attached to the pelvis. A deep learning model based on Convolutional Neural Network (CNN) was developed and trained using approximately 16 min of data collected from a single subject. Ground truth data for training was obtained from an IMU-based motion capture system. The model's accuracy was evaluated using the Mean Absolute Error (MAE) $\pm$ Standard Deviation (SD), averaged across all estimated joint angles, resulting in 2.0 $\pm$ 2.1 degrees. Furthermore, the trained model was implemented in an iPad application, enabling real-time estimation and visualization as a 2D animation of the lower body. This implementation demonstrates potential for extensive utilization in rehabilitation settings and sport science, representing a significant step towards realizing a low-cost, location-independent, and portable gait analysis solution.

Keywords: Gait analysis · real-time estimation · joint angle estimation · wearable sensor · IMU · deep learning

1 Introduction

Gait analysis is a quantitative tool for analyzing human walking patterns, widely used in clinical medicine and rehabilitation. This method enables objective and detailed evaluation of gait pattern, playing a crucial role in the diagnosis of various diseases and disorders, treatment planning and decision, and the assessment of treatment efficacy [1, 2]. Gait parameters include critical measurements such as velocity, step length, gait cycle, ground reaction force (GRF), and joint angles. These parameters are closely associated with various gait disorders, including neurodegenerative disease like Parkinson's disease [3]. The analysis of these parameters allows for a comprehensive understanding and interpretation of various aspects of gait.

© The Author(s), under exclusive license to Springer Nature Singapore Pte Ltd. 2025
J. Mo and P. Chutima (Eds.): ICMSC 2024, LNME, pp. 65–73, 2025.
https://doi.org/10.1007/978-981-96-5380-5_7

To capture and analyze the complex mechanics of human locomotion, clinicians and researchers employ a variety of methods [4]. These approaches can be broadly categorized into two main groups: wearable sensor-based methods and non-wearable sensor-based methods. Wearable sensor-based methods primarily include insole pressure sensor system, inertial measurement unit (IMU) system, electromyography (EMG) system [5]. Non-wearable sensor system primarily includes optical motion capture system, floor sensors, image processing techniques.

Gait analysis using wearable sensors has gained significant attention due to its inexpensive, convenient, and efficient nature [6, 7]. While these methods have been used for decades, recent advancements in deep learning and other machine learning techniques have significantly enhanced the ability to analyze and interpret data from these sensors. This has led to increased interest in using a minimal number of wearable sensors for human motion estimation.

Recent studies have demonstrated the potential of this approach. For instance, a neural network-based method was developed to estimate lower-limb joint angles using a single IMU, achieving promising result [8]. Furthermore, a deep learning model using recurrent neural networks demonstrate the ability to estimate whole-body joint angles during walking using only three IMUs placed on the left wrist and both heels, achieving promising accuracy [9]. Utilizing two insole sensors on each foot, a deep learning approach successfully estimated lower limb joint angles [10]. A combination of IMU and EMG data with Random Forest (RF) or Convolutional Neural Network (CNN) models demonstrated superior performance in estimating lower-limb kinematics, kinetics, and muscle forces, suggesting a promising alternative to traditional optical motion capture for 3D gait analysis [11].

In this research, we present a system that estimate leg joint angles from a single IMU attached to the pelvis. For this estimation, we employ a deep learning algorithm, specifically a CNN. The ground truth data used for training is obtained from an IMU-based motion capture system. Furthermore, this research aims to estimate joint angles in real-time and display in a 2D human animation. This provides immediate feedback, offering practical benefits. The ability to instantly visualize gait patterns could enhance the assessment and treatment processes in clinical settings. This approach has the potential to significantly contribute to the realization of ubiquitous healthcare in daily life, extending beyond its application in medical and rehabilitation settings.

2 Experimental Setup and Data Collection

2.1 Experimental Setup

This study utilized a comparatively low-cost Inertial Measurement Unit (IMU) sensor (AHRS EBIMU-9DOFV5-R3, E2BOX) to estimate leg joint angles. E2BOX IMU is equipped with triaxial accelerometers and gyroscopes. These sensors measure 6-axis motion data: three-dimensional acceleration and three-dimensional angular velocity. For ground truth data acquisition, we employed an IMU-based motion capture system (Xsens MTw Awinda, Movella Inc.). This system uses wireless inertial-magnetic sensors with advanced algorithms to provide accurate 3D kinematic data in various environments [12].

Both sensor systems were configured to collect data at a sampling rate of 100 Hz. A preliminary test confirmed that the timestamp discrepancies between the two sensors were negligible, ensuring that timing misalignments would not significantly impact the data unless continuous collection occurred over an exceptionally extended period.

The sensor placement is illustrated in Fig. 1. The Xsens system utilized seven sensors to capture lower body movements. These sensors were positioned on the subject's body: one posterior pelvis, one on the lateral aspect of each thigh, one on each shin, and one on each foot. The E2BOX sensor was affixed adjacent to the pelvic Xsens sensor.

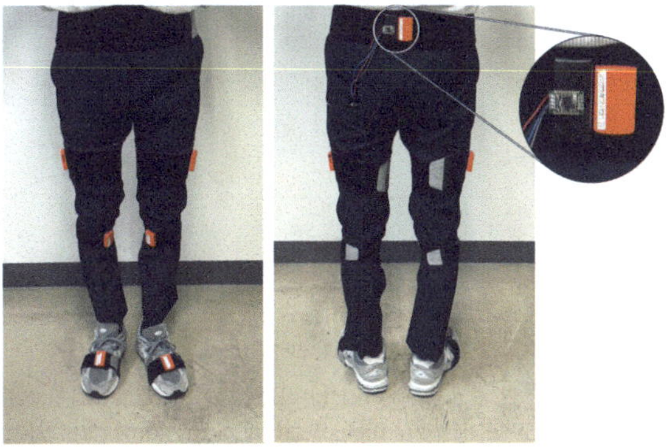

Fig. 1. Experimental setup. Seven Xsens sensors (orange) were attached to the lower body. E2BOX sensor was placed adjacent to the pelvic Xsens sensor.

2.2 Data Collection

The training data was collected from a single subject, a healthy 22-year-old adult male. The data collection process lasted approximately 16 min. To include more diversity and avoid early overfitting, the subject was instructed to walk different speeds: fast, normal, and slow, allocating approximately 5 min to each speed. Additionally, the subject maintained a standing posture for 1 min to capture static balance.

To ensure accurate synchronization between the E2BOX IMU sensor and Xsens Motion capture system data, we employed a physical synchronization method. At the beginning and end of the experiment, both sensors attached to the pelvis were lightly tapped simultaneously. This action produced a distinct acceleration spike in both sensor data, which served as a clear synchronization point for data alignment process.

Figure 2 presents the synchronized data from the E2BOX IMU and the joint angle data obtained from Xsens motion capture system. The IMU data exhibits complex patterns that are challenging for human interpretation, in contrast to the more readily understandable periodic changes in joint angles. While these complex IMU signals are difficult for humans to directly correlate with joint movements, our research aims to estimate joint angles from this intricate IMU data.

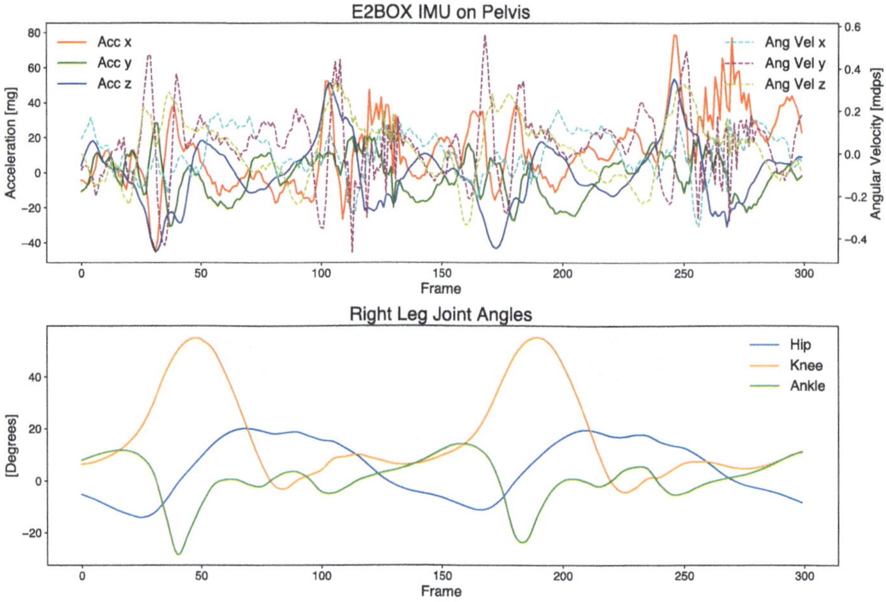

Fig. 2. Synchronized E2BOX IMU data and joint angle data during gait. Acceleration is measured in mg (milligravity), and angular velocity is measured in mdps (millidegrees per second).

2.3 Dataset Preparation

After synchronizing the IMU data from E2BOX IMU and joint angle data from Xsens, dataset for training was created. The collected data was first categorized into four types: three different walking speed (fast, normal and slow) and standing posture. We then split this data into training, validation, and test sets using an 8:1:1 ratio, respectively.

We employed sliding window approach to segment the data for each category. For the training and validation sets, a window size of 100×6 (100 represents one second of data, 6 represents data from IMU) moved along the time series with step size of 10 (represents 0.1 s). For the test set, we used the same window size but reduced the step size to 1, allowing for predictions at every time step. For each window, the corresponding ground truth data was a 1×6 vector, representing hip, knee, and ankle joint angles of both legs. This approach allows the model to learn to estimate the current joint angles using the preceding one second of IMU data.

After creating dataset for each category, we combined them to form our final dataset. This process yielded 7,566 samples for training, and 947 samples for validation. For the test set, using a step size of 1 resulted in 9,460 samples. Each input sample consists of 100 time steps with 6 features, while each output sample is a vector of 6 joint angles.

3 Estimation Algorithm

3.1 Model Architecture

In this study, Convolutional Neural Network (CNN) was employed for joint angle estimation. CNN are widely used deep learning architecture for various tasks, including image processing, speech recognition, and time series analysis. They are characterized by their ability to learn local features effectively through convolutional layers. For tasks involving learning relationships between IMU sensor data and joint angles, CNNs are well-suited due to their ability to capture temporal patterns in the data. Moreover, the efficient processing structure of CNNs can allow for fast computation, making them suitable for applications requiring real-time performance.

The CNN model developed for this research consists of three 1-dimensional convolutional layers and four dense layers. The input shape is (100, 6), representing 100 time steps of 6 dimensional IMU data. The convolutional layers use 256, 128, and 64 filters with kernel sizes of 3, 5, and 7 respectively, each followed by leaky ReLU activation function. Max pooling is applied after the first two convolutional layers. The subsequent dense layers contain 64, 32, and 16 neurons, with an output layers of 6 neurons that predict the joint angles. Leaky ReLU activation is used throughout, except for the output layer.

3.2 Training Process

The model was trained for 100 epochs with a batch size of 64. Adam optimizer was used as the optimization algorithm and Mean Absolute Error (MAE) was chosen as the loss function. The dataset preparation is described in detail in Sect. 2.3. Figure 3 shows the progression of the loss function, which demonstrates steady convergence.

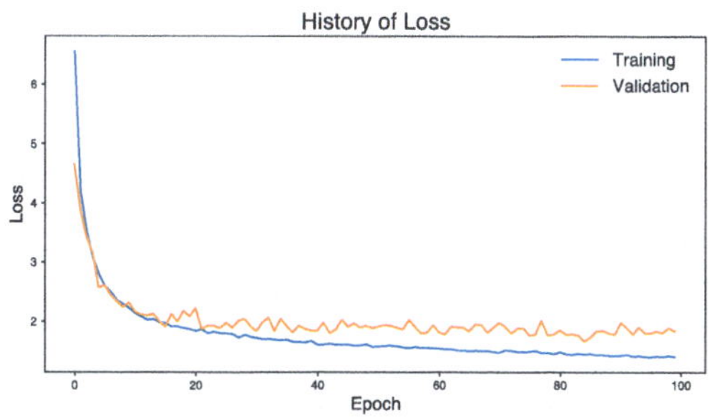

Fig. 3. Training and validation loss history

3.3 Model Performance and Evaluation

The trained model performance was evaluated using a separate test dataset. We used Mean Absolute Error (MAE), and Standard Deviation (SD) as the evaluation metrics.

The average MAE $\pm$ SD across all joint angles was $2.0 \pm 2.1°$, indicating that the model can estimate joint angles with high accuracy using only a single IMU attached to the pelvis (see Table 1). Figure 4 compares the ground truth joint angles with the estimated joint angles during walking. While the model doesn't perfectly capture every nuance of complex movements, it closely tracks the overall trajectory of joint angles.

Table 1. MAE $\pm$ SD of joint angle estimation using a single IMU on pelvis (in degrees)

Left			Right			Average
Hip	Knee	Ankle	Hip	Knee	Ankle	
2.1 ± 2.1	2.1 ± 2.4	2.1 ± 1.8	1.8 ± 1.5	2.2 ± 2.5	1.8 ± 1.9	2.0 ± 2.1

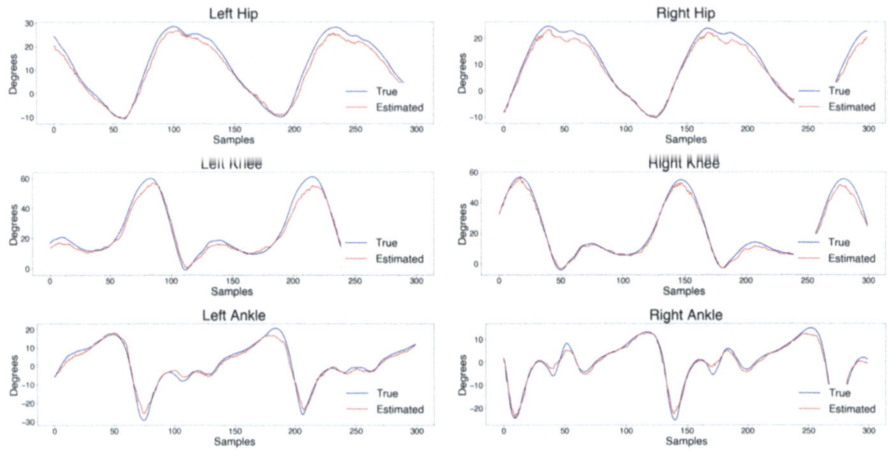

Fig. 4. Comparison of estimated joint angles with ground truth measurements

4 Real-Time Estimation and Visualization

4.1 System Architecture for Real-Time Implementation

Figure 5 shows the system configuration diagram of our real-time implementation for joint angle estimation from IMU data. This system comprises three primary elements: an ESP32 microcontroller, an E2BOX IMU sensor, and a tablet device (iPad). The ESP32 and E2BOX IMU are connected via Universal Asynchronous Receiver Transmitter (UART) communication, with the ESP32 reading acceleration and angular velocity from the E2BOX IMU at a rate of 100 Hz. The ESP32 microcontroller transmits this acquired data to the tablet device via wireless communication of Bluetooth Low Energy (BLE). The ESP 32 is powered by a mobile battery, and it in turn supplies power to the E2BOX IMU, creating a self-contained sensing unit.

The iPad application was developed using Flutter, a cross-platform development framework created by Google. Flutter was primarily chosen for its potential to easily extend the application to other platforms in the future, while maintaining a single codebase. To enable inference on the iPad, the trained model was converted to TensorFlow Lite format. When data is received from the ESP32 microcontroller, it is reshaped into a 100×6 shape to match the model's input requirements. This formatted data is then input into the model, which outputs estimated joint angles. These estimated angles are subsequently visualized as a 2D animation of the lower body in the sagittal plane. This process allows for representation of the subject's walking motion, translating raw IMU data into lower body kinematics in real-time.

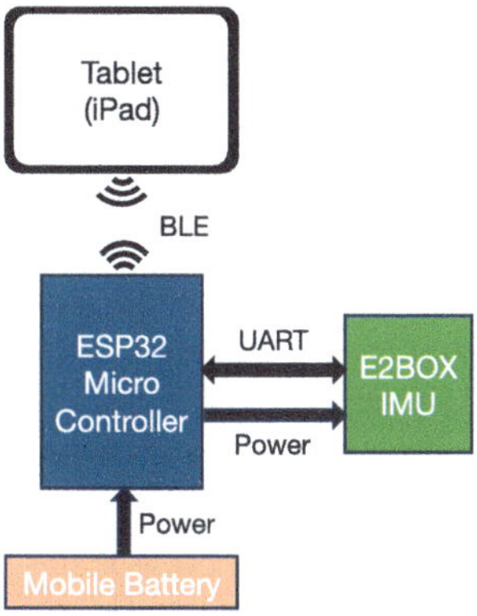

Fig. 5. System configuration diagram of the real-time implementation for joint angle estimation using IMU data

4.2 Implementation Result

Figure 6 represents the comparison between the real-time 2D animation of estimated joint angles and the corresponding sequential photographs of the subject's walking. This

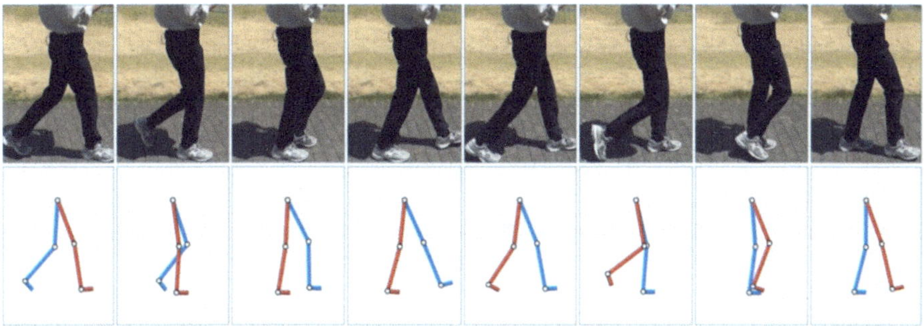

Fig. 6. Comparison of one gait cycle between sequential photo (top row) and corresponding real-time 2D animation of estimated joint angles (bottom row).

comparison clearly demonstrates that the estimated joint angles of the hip, knee, and ankle closely match the actual gait pattern. The inference time on iPad ranged from 1 to 2 ms.

5 Discussion and Future Work

This study demonstrated the feasibility of approach for real-time joint angle estimation using a single IMU attached to the pelvis. The developed system, comprising a relatively inexpensive IMU and an iPad application showed promising result. The model achieved an overall average MAE $\pm$ SD of $2.0 \pm 2.1^\circ$ across all joint angles. These results demonstrate the potential for simplifying gait analysis in clinical and rehabilitation setting, potentially reducing patient burden. This portable setup has the potential to allow for gait monitoring in everyday environments. However, this study was conducted with a single subject, which limits the generalizability of our findings.

Future research directions should focus on expanding the system's capabilities. Firstly, increasing the diversity of the subject and improving model generalizability and ensuring robust performance. Secondly, estimating additional gait parameters such as step length, cadence, and swing/stance time would enhance the system's utility in comprehensive gait analysis. Lastly, conducting practical applications with specific groups, such as elderly individuals or pre/post-hip surgery patients, would provide valuable insights into the system's efficacy in real-world clinical and rehabilitation settings.

Acknowledgement. This research was supported by JSPS KAKENHI Grant Number JP22K04012.

References

1. Baker, R., Esquenazi, A., Benedetti, M.G., Desloovere, K.: Gait analysis: clinical facts. Eur. J. Phys. Rehabil. Med. **52**, 560–574 (2016)
2. Baker, R.: Gait analysis methods in rehabilitation. J. Neuroeng. Rehabil. **3**, 4 (2006)
3. Zanardi, A.P.J., et al.: Gait parameters of Parkinson's disease compared with healthy controls: a systematic review and meta-analysis. Sci. Rep. **11**, 752 (2021)
4. Muro-de-la-Herran, A., Garcia-Zapirain, B., Mendez-Zorrilla, A.: Gait analysis methods: an overview of wearable and non-wearable systems, highlighting clinical applications. Sensors **14**, 3362–3394 (2014)
5. Chen, S., Lach, J., Lo, B., Yang, G.-Z.: Toward pervasive gait analysis with wearable sensors: a systematic review. IEEE J. Biomed. Health Inform. **20**, 1521–1537 (2016)
6. Tao, W., Liu, T., Zheng, R., Feng, H.: Gait analysis using wearable sensors. Sensors **12**, 2255–2283 (2012)
7. Washabaugh, E.P., Kalyanaraman, T., Adamczyk, P.G., Claflin, E.S., Krishnan, C.: Validity and repeatability of inertial measurement units for measuring gait parameters. Gait Posture **55**, 87–93 (2017)
8. Alemayoh, T.T., Lee, J.H., Okamoto, S.: Leg-joint angle estimation from a single inertial sensor attached to various lower-body links during walking motion. Appl. Sci. **13**, 4794 (2023)

9. Kumano, Y., Kanoga, S., Yamamoto, M., Takemura, H., Tada, M.: Estimating whole-body walking motion from inertial measurement units at wrist and heels using deep learning. Int. J. Automot. Technol. **17**, 217–225 (2023)
10. Alemayoh, T.T., Lee, J.H., Okamoto, S.: A neural network-based lower extremity joint angle estimation from insole data. In: 2023 20th International Conference on Ubiquitous Robots (UR), pp. 787–791. IEEE (2023)
11. Moghadam, S.M., Yeung, T., Choisne, J.: A comparison of machine learning models' accuracy in predicting lower-limb joints' kinematics, kinetics, and muscle forces from wearable sensors. Sci. Rep. **13**, 5046 (2023)
12. Schepers, M., Giuberti, M., Bellusci, G.: Xsens MVN: consistent tracking of human motion using inertial sensing. Xsens Technol. **1**(8), 1–8 (2018)

Micro Mechanical System Design

Optimization of Automatic Boar Recognition Models for Microcontrollers by Framework Selection and Quantization Methods

Ryunosuke Niimoto, Francis Ifaso Besala, Jae Hoon Lee(✉), and Shingo Okamoto

Graduate School of Science and Engineering, Ehime University, 3 Bunkyocho, Matsuyama, Ehime 790-0826, Japan
lee.jaehoon.mc@ehime-u.ac.jp

Abstract. In Japan, damage to crops by wild boars and other wild animals is becoming increasingly serious due to the aging of hunters and the lack of effective trapping devices, causing significant damage to farmers. As a countermeasure, farmers are using traps such abox type trap, fence trap and wire trap, however these methods are essentially manual thus limited in efficiency to responds against the crop harmful animals. Therefore, a recognition algorithm is needed for recognizing wild boar presence and capture them automatically by Artificial intelligence. Therefore, in our previous study, a recognition algorithm was used to develop to recognize boars in box traps and a digitally controllable capturing device, and these were introduced into a microcontroller; however, there were limitations in accuracy and processing efficiency. Therefore, the objective of this study is to select the best AI model in terms of accuracy and speed on a microcontroller. A new dataset was created using data collected in the experimental site, and AI models were trained using two different frameworks: Pytorch for flexible processing and Darknet for lightweight and fast inference. The data types were then converted using quantization techniques and implemented on a microcontroller to compare and verify each data type and framework. This comparison allowed us to select the optimal AI model on the microcontroller.

Keywords: Wild boars · YOLO · Microcontroller · Pytorch · Darknet · Quantization

1 Introduction

In recent years, the damage of crops harmful animal damage such as wild boars, deer, and monkeys has become increasingly serious in Japan. Wild boars cause tremendous damage to farmers due to their behavior as feeding on a variety of crops, digging ground for searching food, trampling further damages plants. According to a survey conducted by the Ministry of Agriculture, Forestry and Fisheries in 2022, the amount of damage to crops caused by wild animals reached approximately 15.6 billion yen, with wild boars, deer, and monkeys accounting for approximately 70% of this total [1]. To deal with this

© The Author(s), under exclusive license to Springer Nature Singapore Pte Ltd. 2025
J. Mo and P. Chutima (Eds.): ICMSC 2024, LNME, pp. 77–87, 2025.
https://doi.org/10.1007/978-981-96-5380-5_8

situation, the national and local governments are taking various measures, However, most of hunters are aging people and also most of trap use are manual thus lack of effect. Many of the automatic trapping devices on the market are big and expensive, basically designed for research institutions or big companies. Therefore, it is difficult for ordinary farmers to use them. Then, farmers utilize generally box trap as it is easy of usage, but a normal box trap cannot respond to snare a specific harmful animal. Therefore, an intelligent system, low cost is needed for capturing harmful animal automatically. In our previous research, recognition algorithm was developed and embedded in microcontroller. Furthermore, an automatic trigger activated by a digital signal controlled by the AMB82-mini micro controller was implemented [2]. In this study, the yolov7-tiny model, which was used previously, has trained using two different frameworks. The Darknet framework and the Pytorch framework with a upgraded dataset including new data collected in the experimental area and data used in our previous research. The trained models were modified in terms of data type using quantization techniques in order to reduce the weight of each model and embedded on AMB82-mini microcontroller [3]. The selection of the optimal setting for implementation on a microcontroller a performance test was made according to balance detection accuracy and processing speed, four different settings are used: data types UINT8 and INT16 for the Darknet framework, data types UINT8 and INT16 for the Pytorch framework. This research will be used to select the optimal AI model on the microcontroller, aiming to improve both accuracy and speed and processing efficiency for use in the real site application.

2 Dataset Preparation

Training in YOLO requires a dataset consisting of images and their corresponding correct answer labels [4]. The correct labels must contain the number meaning the class of the object and the bounding box coordinate of detected object in text format. Therefore, a dataset containing images collected from trail cameras and the internet in our previous studies and the correct labels attached to them was used. In addition to the dataset, additional image data is obtained from trail cameras and data collection systems installed in the field for machine learning. The annotation process was carried out using Microsoft Corporation's VoTT 1.19 application software to produce correct labels for the collected image data [5]. A dataset was created with 2046 images for training, 200 for validation and 163 for testing, together with a dataset from a previous study with labelling according to this format. The dataset contains a total of 2409 data including with 429 images of boars in the box trap, 500 images of boars outside the box trap, 471 images of raccoons in the box trap, 491 images of raccoons outside the box trap and 518 images of empty box traps, each with a class name were trained as (a) Boar, (b) Boar in trap, (c) Raccoon, (d) Raccoon in trap and (e) Nothing. Examples of each class are shown in Fig. 1.

(a) Boar (b) Boar in trap

(c) Raccoon (d) Raccoon in trap

(e) Nothing

Fig. 1. Examples of Training Image Data

3 Training Process Using the Pytorch Framework

3.1 Training Conditions

Pytorch is one of the open source neural network frameworks used primarily in machine learning and deep learning research and development [6]. The framework is flexible and easy to use and features a dynamic computational graph. This makes it easy to debug and allows for rapid experimentation and model tuning, a major advantage. Pytorch can be used with pre-trained pt files and yaml files for many different types of tasks. In this study, yolov7-tiny.yaml was used as the CNN configuration file because real-time processing using a microcontroller is assumed. We modified yolov7-tiny.yaml based on the information provided by the developer. First, change the number of classes in the yolo layer, the classifier, from the default 80 to 5. The class name was set to be the same as the correct answer label used for the dataset. The resize, batch size, and number of epochs were then set as shown in the Table 1, and training was performed using commands specified by the developer on a desktop computer with the specifications shown in Table 2.

Table 1. Training condition

Condition	Setting value
Resize width	640
Resize height	640
Batch size	16
epochs	100

Table 2. Specifications of desktop computer for training

Part	Product name
Processor	Intel® Core™ i7-7700KF CPU @ 3.60 GHz
Memory	15 GB
Operating system	Ubuntu 22.04.4 LTS
GPU	NVIDIA GeForce GTX1080Ti

3.2 Training Result

Figure 2 shows the change in the loss function due to learning. The loss function has converged sufficiently, indicating that learning has been done well without over-learning. Figure 3 shows the correct response rate the weights after 100 learning cycles show that the training and validation images have a correct response rate of about 93%. In addition, for the specific classes "boar" and "boar in trap," the accuracy is high, more than 98% for both.

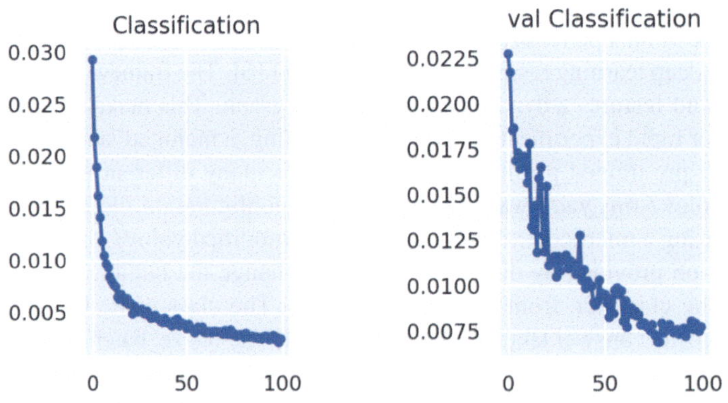

Fig. 2. Validation and Training for Loss Function

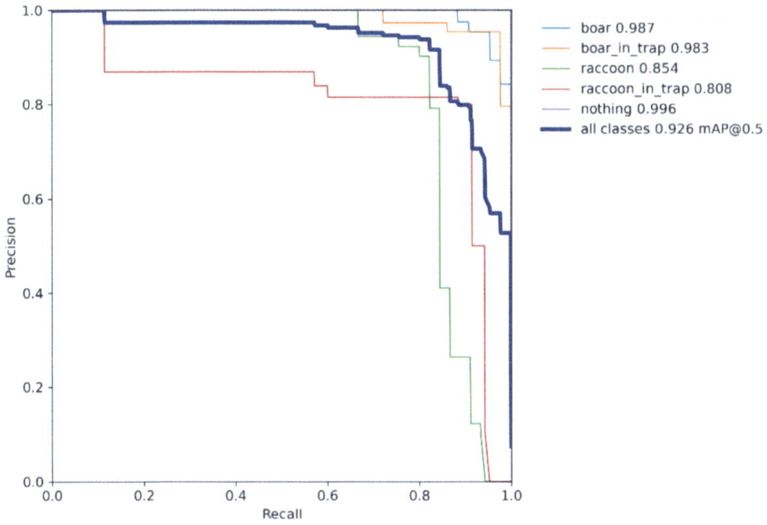

Fig. 3. Training Results (Pytorch)

4 Training Process Using the Darknet Framework

4.1 Training Condition

Darknet is one of the open source neural network frameworks mainly used in YOLO object detection systems [7]. The framework has high performance in real-time object detection, especially in quickly and accurately identifying objects in an image by analyzing the entire image at once. Darknet can identify various types of objects. The framework is written in C and is lightweight, making it easy to implement on many devices and systems. Change yolov7-tiny.cfg to be specified by the developer. Change the number of classes in the yolo layer, which is the classifier, from the default 80 to 5. Change the number of filters in the convolutional layer, which is one before the yolo layer, based on the following formula.

$$N_f = (N_c + 5) \times 3$$

N_f: Number of Filters
N_c: Number of Classes

The number of classes is 5, so the number of filters is 30. Yolov7-tiny.cfg contains two yolo layers, so these changes were made to all of them.

In yolov7-tiny.cfg, the image resizes, training frequency, batch size, etc. are specified. In this study, the settings were made to match the Pytorch framework as closely as possible, as shown in Table 3. Then, training was performed using the commands specified by the developer on a desktop computer with the specifications shown in Table 2, which was used in Sect. 3.

Table 3. Training condition

Condition	Setting value
Resize width	640
Resize height	640
Iterations	12787
Batch size	16
Momentum	0.9
Burn in	1000
Learning rate	0.0261

4.2 Training Result

Figure 4 illustrates the training outcomes. The red area in the figure depicts the mean average precision (mAP), which is an overall average index that encompasses both the model's prediction accuracy and its localization ability. The model achieved an accuracy rate of 90.0%, indicating that it is capable of accurately recognizing and locating the object. The blue area represents the value of the loss function. As training progresses, the

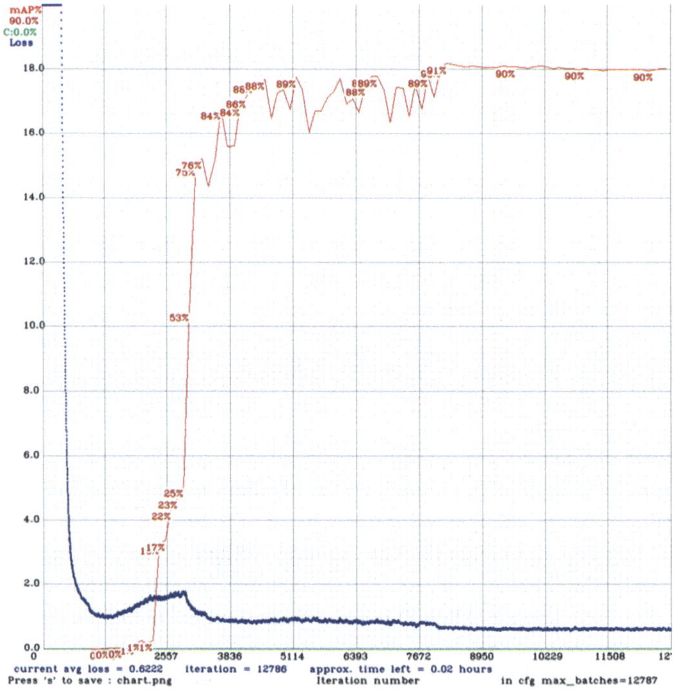

Fig. 4. Training Result (Darknet)

graph of the loss function shows a decrease in value, ultimately converging to indicate that the model has been trained appropriately without over-training.

5 Comparative Verification Using Microcontrollers

5.1 About AMB82-Mini Microcontrollers

The AMB82-mini, which uses Realtek's RTL8735BDM SoC, is a development board specialized for AI and IoT applications [8]. Figure 5 shows the AMB82-mini. The board is optimized for battery-powered devices and AI network camera applications, specializing in power efficiency and enhanced IoT security. Key features include Wi-Fi and Bluetooth support, H264/H265 video encoding, 1080p sensor, various I/O options, and the ability to use AI models. Support for the Arduino IDE allows users to create programs and upload them to the board using the Arduino development environment. The Table 4 below lists the specifications of this microcontroller.

Fig. 5. AMB82-mini microcontrollers

Table 4. Specifications of AMB82-mini

Part	Product name
Processor	ARM Cortex-M33 (RTL8735BDM SoC)
Memory	512 KB SRAM, 128 KB ROM, 16 MB Flash
Operating system	FreeRTOS, Arduino IDE
Accelerator	Built-in Neural Processor Unit (NPU)

5.2 Weight Reduction by Quantization

Importing the weight files trained on the AMB82-mini as they are a heavy load for the microcontroller and difficult to use directly. However, by utilizing an online tool provided by the company that developed the AMB82-mini, a quantized binary file is automatically generated, which can then be used to import the weight file into the AMB82-mini. Originally, the data type used by YOLO is Float32, but quantization converts the data type to Int8 or Int16. This conversion reduces data weight and greatly increases processing speed, allowing for more efficient real-time processing.

5.3 Verification Method

In this experiment, 163 images data for testing will be used for verification. An overview of the verification environment is shown in Fig. 6. The AMB82-mini performs recognition based on the image data acquired from the camera module and sends the results to the PC via Wi-Fi. The AMB82-mini recognizes the image data acquired from the camera module and sends the results to the PC via Wi-Fi. The recognition accuracy is evaluated based on the output results displayed on the PC. Furthermore, the processing speed is acquired via serial communication and checked on a serial monitor. These procedures are verified with four different frameworks and quantization settings: data types UINT8 and INT16 for the Darknet framework and data types UINT8 and INT16 for the Pytorch framework, and their performance is compared.

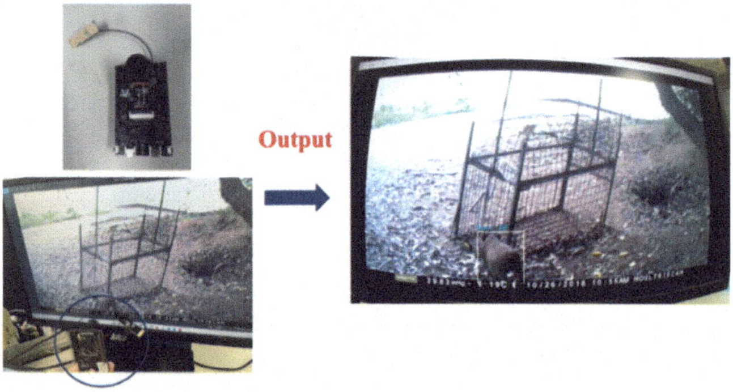

Fig. 6. Overview of Validation Environment

5.4 Verification Results

In this experiment, detection accuracy per class and inference speed per image were compared for two frameworks, Pytorch and Darknet, based on INT16 and UINT8 quantization settings. The results are shown in the Tables 5, 6. Overall, Pytorch showed higher detection accuracy than Darknet. In particular, in the "Raccoon" and "Raccoon in trap" classes, Pytorch's INT16 setting achieved 90.0% and 85.7% accuracy, respectively, significantly better than Darknet's accuracy (82.2% and 62.0%). Darknet, on the other hand, achieved an accuracy of 93.3% for both INT16 and UINT8 in the "Boar in trap" class, showing stable results that are not affected by quantization. However, as shown in the Fig. 7, the "Raccoon in a trap" class was frequently misidentified as "Boar in a trap," confirming that the accuracy was low. As for inference speed, Darknet was faster than Pytorch. In particular, in the UINT8 quantization setting, Darknet completed inference on a single image in 0.19 s, slightly faster than Pytorch (0.20 s). The effect of quantization size was examined. Regarding the effect of quantization size, INT16 showed higher detection accuracy than UINT8. For example, in the "Boar" class, Pytorch's INT16 setting achieved 85.1% accuracy, better than UINT8's 80.8%. In the "Nothing" class, Pytorch's INT16 setting achieved 98.0% accuracy compared to 93.0% for UINT8. On the other hand, UINT8, with its smaller quantization size, was noticeably faster than INT16 in inference speed, 0.43 s for INT16 on Pytorch versus 0.20 s for UINT8, and 0.39 s for INT16 versus 0.19 s for UINT8 on Darknet. UINT8 showed a 2-fold speed improvement in both frameworks. The results of this study suggest that the UINT8 setting in the Pytorch framework is particularly superior when considering the balance between accuracy and processing speed. The ability to increase inference speed while maintaining accuracy makes Pytorch's UINT8 setting the best choice for real-time processing and systems with resource constraints.

Table 5. Verification result of Accuracy per Class [%]

Class	Pytorch INT16	Pytorch UINT8	Darknet INT16	Darknet UINT8
Boar	85.1	80.8	83.3	74.8
Boar in trap	90.7	90.7	93.3	93.3
Raccoon	90.0	90.0	82.2	73.0
Raccoon in trap	85.7	73.0	62.0	58.9
Nothing	98.0	93.0	95.0	93.0

Table 6. Verification result of processing speed

	Pytorch INT16	Pytorch UINT8	Darknet INT16	Darknet UINT8
Speed (s)	0.43	0.20	0.39	0.19

Fig. 7. Example of false detection

6 Conclusions

The goal of this study was to develop an optimal artificial intelligence for wild boar detection on microcontrollers, Therefore, the detection accuracy and inference speed using INT16 and UINT8 quantization settings in the Pytorch and Darknet frameworks using the YOLO algorithm were compared and verified. The results revealed that Pytorch's UINT8 setting is the best choice for resource-constrained environments requiring real-time processing, offering an excellent balance between accuracy and processing speed. In particular, it was confirmed that it is possible to improve processing speed while maintaining accuracy, and the foundation is now in place for future practical applications. As for future issues, we will improve recognition accuracy and system stability through further collection of local data and expansion of the data set. In addition, we will develop a stand-alone automatic boar trapping system that uses solar panels, batteries and will improve its usefulness through installation in the field, with the aim of commercializing it.

Acknowledgement. This research was supported by JSPS KAKENHI Grant Number JP22K04012.

References

1. Ministry of Agriculture, Forestry and Fisheries. The current state and countermeasures of wildlife damage (2024). https://www.maff.go.jp/j/seisan/tyozyu/higai/attach/pdf/240605-18.pdf. Accessed 06 Oct 2024
2. Niimoto, R., Ifaso, F.B., Lee, J.H., Okamoto, S.: Automatic recognition of wild boar using deep learning. In: Proceedings of SICE Festival 2024 with Annual Conference, pp. 580–583 (2024)
3. Zhao, T., Cui, Z., Xie, F., Yang, M.: Quantized YOLO for resource-constrained edge devices. In: 2020 IEEE International Conference on Consumer Electronics (ICCE), pp. 1–4 (2020)
4. Redmon, J., Farhadi, A.: YOLOv3: an incremental improvement. arXiv preprint arXiv:1804.02767 (2018)
5. Microsoft "visual object tagging tool (VoTT): open-source annotation and labeling tool" (2020). https://github.com/microsoft/VoTT. Accessed 11 Sept 2024

6. Bochkovskiy, A., Wang, C., Liao, H.: YOLOv4: optimal speed and accuracy of object detection. arXiv preprint arXiv:2004.10934 (2020)
7. Redmon, J., Farhadi, A.: YOLO: you only look once. In: Proceedings of the IEEE Conference on Computer Vision and Pattern Recognition (CVPR), pp. 779–788 (2016)
8. Realtek IoT/Wi-Fi MCU Solutions, AmebaPro2 Boards. https://www.amebaiot.com/en/amebapro2/. Accessed 07 Oct 2024

Computational Fluid Dynamics of Microplastics Separation in Hydrocyclones: Effects of Operating Parameters

Dulyapat Thiemsakul[1], Ratchanon Piemjaiswang[2], Teerawat Sema[1,3], Pornpote Piumsomboon[1,3], and Benjapon Chalermsinsuwan[1,2,4](✉)

[1] Fuels Research Center, Department of Chemical Technology, Faculty of Science, Chulalongkorn University, Bangkok, Thailand
{teerawat.se,pornpote.p,benjapon.c}@chula.ac.th
[2] Environmental Research Institute, Chulalongkorn University, Bangkok, Thailand
ratchanon.p@chula.ac.th
[3] Center of Excellence on Petrochemical and Materials Technology, Chulalongkorn University, Bangkok, Thailand
[4] Advanced Computational Fluid Dynamics Research Unit, Chulalongkorn University, Bangkok, Thailand

Abstract. Microplastics have emerged as a significant environmental concern due to their infiltration into freshwater and marine ecosystems. To address this challenge, the hydrocyclone has been investigated as a potential solution for separating microplastics from water. Utilizing strong centrifugal forces, the hydrocyclone effectively separates discrete phases, such as microplastics, from continuous phases. Furthermore, hydrocyclones offer flexible operation, high capacity, and low operation and maintenance costs. In this study, computational fluid dynamics (CFD) modelling was used to investigate the hydrodynamics and microplastic separation in hydrocyclones with different operating parameters to separate microplastics from water. From the computational result, microplastics were separated with 83.06% plastics recovery in case inlet velocity of 10 m/s and PET microplastics concentration of 0.1%vol. In addition, the ANOVA demonstrated that microplastic density and inlet feed velocity were significant factors affecting microplastic separation in the hydrocyclone.

Keywords: Microplastics · Computational fluid dynamics · Hydrocyclones

1 Introduction

Microplastics have emerged as a significant environmental concern problem due to the consistently increasing annual global demand for plastics in recent years. Microplastics are tiny plastic pieces smaller than 5 mm that can end up in freshwater and marine environments, harming both nature and public health [1, 2]. Microplastics can act as primary pollutants, coming directly from products or materials that contain tiny plastic particles like microbeads, microfibers, and plastic pellets [3]. Because of their small size, removing microplastics from water bodies, sediments, and air using current methods is extremely challenging.

© The Author(s), under exclusive license to Springer Nature Singapore Pte Ltd. 2025
J. Mo and P. Chutima (Eds.): ICMSC 2024, LNME, pp. 88–96, 2025.
https://doi.org/10.1007/978-981-96-5380-5_9

The hydrocyclone is a device designed to separate particles based on size or density, leveraging their physical characteristics. It generates a strong centrifugal force that effectively separates discrete phases from continuous ones [4]. Hydrocyclones are commonly used in industries like mining, chemicals, petroleum, and biotechnology due to their operational flexibility, high capacity, and low maintenance costs [5, 6]. The separation process begins with the tangential entry of a pressurized fluid and solid mixture into the cylindrical chamber, creating a vortex. As the vortex moves downward through the conical section, heavier solids are pushed to the sides by centripetal force, slow down due to drag forces, and eventually slide to the underflow for ejection. Meanwhile, lighter particles and liquids form a secondary vortex at the center, rising and exiting through the overflow via the vortex finder.

The hydrocyclone is a device designed to separate particles based on size or density, leveraging their physical characteristics. It generates a strong centrifugal force that effectively separates discrete phases from continuous ones [7]. Hydrocyclones are commonly used in industries like mining, chemicals, petroleum, and biotechnology due to their operational flexibility, high capacity, and low maintenance costs. The separation process begins with the tangential entry of a pressurized fluid and solid mixture into the cylindrical chamber, creating a vortex. As the vortex moves downward through the conical section, heavier solids are pushed to the sides by centripetal force, slow down due to drag forces, and eventually slide to the underflow for ejection [8]. Meanwhile, lighter particles and liquids form a secondary vortex at the center, rising and exiting through the overflow via the vortex finder.

To better understand the complex flow dynamics inside hydrocyclones, computational fluid dynamics (CFD) simulations have been employed [9]. CFD modeling, particularly with the ANSYS Fluent software using a three-dimensional Eulerian-Eulerian approach and granular flow kinetics, was utilized to study liquid-solid flow behavior. This approach offers a detailed analysis of local phase velocities and holdups, which are challenging to measure experimentally. CFD provides a cost-effective and efficient method for analyzing, designing, and optimizing hydrocyclone performance in multiphase systems [10]. In addition to CFD, analysis of variance (ANOVA) was used to statistically evaluate the impact of various parameters on microplastic separation. Parameters such as the type of microplastics, feed velocity, and plastic concentration were examined to assess their significance in the separation process. The combination of CFD and ANOVA offers a robust methodology for enhancing hydrocyclone design and performance.

2 Literature Review

The general methods for removing insoluble solids such as microplastics include flocculation setting, air flotation, rapid sand filtration, membrane filtration, etc. [11]. Hydrocyclones were also a type of solid-liquid separation process, but they worked based on the principles of centrifugal force rather than filtration. Compared to other solid-liquid separation processes, hydrocyclones were efficient and low-maintenance devices that could operate at high pressures and temperatures, making them suitable for a wide range of applications. The flocculation setting is simple to operate and inexpensive. However, the flocculant has poor adsorption effectiveness on fine particles, which reduces the performance of the flocculation setting. Bubble flotation enables continuous operation and can improve separation efficiency [12]. The flotation process is significantly impacted by the size of the bubbles [13]. However, before processing in the industry, the sizes of microplastics are usually unknown, making it challenging to estimate bubble size, which further affects flotation effectiveness. Rapid sand filtration is an efficient and affordable filtration method for filtering fine insoluble particles [14]. However, it has been found that microplastic particles easily attach to silica particles, leading to obstruction and decreased working effectiveness [15].

3 Materials and Methods

This study investigated the hydrodynamics and microplastic separation in hydrocyclones using a three-dimensional Eulerian-Eulerian CFD model implemented in ANSYS Fluent 2022 R2. The liquid and solid phases were modeled mathematically based on the conservation of momentum and mass for each phase [16]. Gidaspow's drag coefficient model, a well-established approach for particle-laden flow systems, was utilized. This model is known for its adaptability to different particle concentrations and fluid properties. A modified version of Gidaspow's drag coefficient correlation was applied to calculate the interphase interactions between the liquid and solid particle phases in the model setup.

The phase-coupled SIMPLE algorithm was employed for pressure-velocity coupling in the simulations, while a first-order upwind scheme was used for spatial discretization. Time-temporal discretization was managed using a pressure-based first-order implicit unsteady solver. Liquid turbulence was accounted for using the k-epsilon (k-ε) turbulence model, which captures the mean flow characteristics in turbulent conditions. Boundary conditions included an inlet-velocity for the entry point and a pressure-outlet for the exit. The simulations were run for 10 s with a time step of 0.0001 s to achieve a steady-state solution.

Figure 1 presents the schematic diagram of the hydrocyclone used in this study. The hydrocyclone features a body diameter of 10 mm, two inlets each 2 mm in diameter, and a spigot with a diameter of 10 mm. The cylindrical section measures 8 mm in length, while the conical section extends 76 mm. The vortex finder is 5 mm long with a thickness of 1 mm. The computational domain was meshed using Fluent's meshing tool, resulting in hexahedral cells with a total count ranging from 40,000 to 140,000.

The computational results were validated against experimental data from He et al. [13], where polymethyl methacrylate (PMMA) particles with a diameter of 10 μm were used. The % recovery of microplastics was evaluated to compare simulation outcomes with experimental findings.

In this study, ANOVA was used to analyze the significance of each parameter on microplastic separation in the hydrocyclone. The studied parameters were the microplastic densities of PS and PET, which are common microplastics in freshwater [17], inlet velocity (6–10 m/s), and microplastic concentration (0.01–0.1%vol) within the range of previous and average microplastic concentration [18, 19]. The uniform diameter of the microplastics was 10 μm, and their densities were 1,090 and 1,250 kg/m^3 for PS and PET, respectively. The variation in particle size may influence the separation performance of mini-hydrocyclones, a factor worth considering in future research endeavors. The microplastics were introduced into the hydrocyclone via water with a density of

998.2 kg/m^3 and a viscosity of 0.001003 kg/m s. Here, temperature fluctuations and the energy equation were not considered. The proposed parameters were varied in different cases to study their effect on microplastic separation in the hydrocyclone (mass of plastics at underflow divided by the total mass of plastics at underflow and overflow). Table 1 shows details of various parameters in this study.

Table 1. Microplastic parameters and the recovery in hydrocyclone with different conditions.

Case	Microplastic type (A)	Inlet velocity [m/s] (B)	Microplastic concentration [%vol] (C)	Microplastic recovery [%]
1	PS	6.0	0.01	63.38%
2	PET	6.0	0.01	72.88%
3	PS	10.0	0.01	66.33%
4	PET	10.0	0.01	77.72%
5	PS	6.0	0.10	62.90%
6	PET	6.0	0.10	75.58%
7	PS	10.0	0.10	66.79%
8	PET	10.0	0.10	83.06%

4 Result and Discussion

4.1 Grid Independency and Time Independency Tests.

The number of computational cells plays a crucial role in achieving accurate numerical results in CFD simulations. However, increasing the number of computational cells significantly raises computational costs [20]. To address this, a grid independence test was conducted to optimize the cell count. In this study, the computational domain was discretized using hexahedral cells to ensure numerical stability and efficiency. For the grid independence test, three configurations with 50,000, 100,000, and 150,000 computational cells were examined. Figure 2 illustrates the % plastics recovery of the hydrocyclone for each configuration. The results indicated consistent numerical outputs for 100,000 and 150,000 cells. Consequently, 100,000 computational cells were selected for the remainder of the study to balance accuracy and computational efficiency.

Additionally, a time independence test was performed to determine the simulation time required for the system to reach a steady-state condition. As shown in Fig. 2, the % plastics recovery stabilized after 3 s of simulation time. Therefore, simulation outputs, including flow rate and solid volume fraction, were averaged after this time point for further comparisons. Moreover, Gidaspow's drag coefficient model was modified to account for its original design based on Geldart group B particles.

The model validation between the experiment and simulation results was performed to confirm the representation of the real system by the developed CFD model. The model

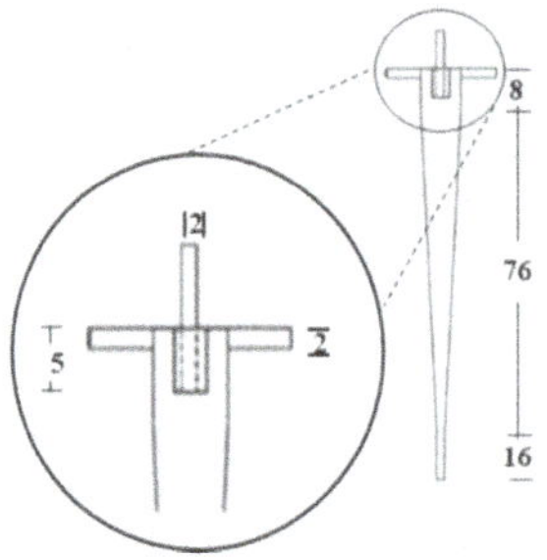

Fig. 1. Schematic diagram of hydrocyclone.

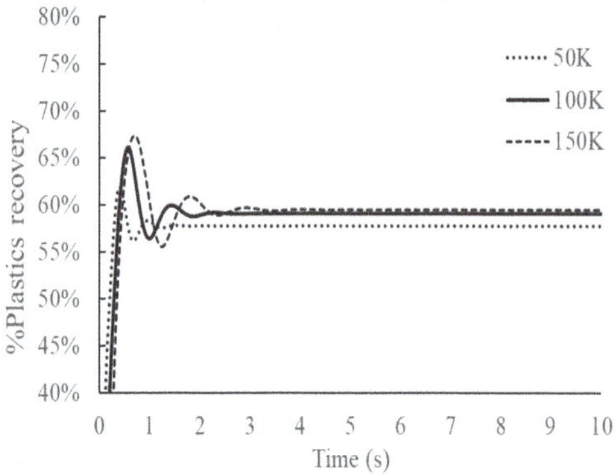

Fig. 2. Plastic recovery with various computational cells.

validation was performed on the same geometry with 0.1% vol of 10 μm polymethyl methacrylate (PMMA) particles and 6.0 m/s of inlet feed velocity. The obtained results from the CFD model of the hydrocyclone were compared with the experimental results of He et al. [13] as shown in Fig. 3. The error bar for the experimental recovery is also shown in Fig. 3 The standard deviation is recorded at 2.99%. These acceptable levels of uncertainty and standard deviations confirm the minimal impact of experimental uncertainty. The obtained results concluded that this study's CFD model with a modified Gidaspow's drag coefficient model could be used to predict the system's behavior in the hydrocyclone.

4.2 Microplastic Separation

The following sections present the results of simulations on important operating parameters of microplastic separation, such as the type of microplastics (A), feed velocity (B), and plastic concentration (C) in the hydrocyclone. The hydrocyclone performance with PS and PET microplastics is also shown in Table 1. Case 8, with a PET inlet velocity of 10 m/s and microplastic concentration of 0.1%vol, showed the highest % recovery, equal to 83.06% while case 5 with a 6 m/s inlet velocity of PS and microplastic concentration of 0.1%vol, had the lowest % recovery, equal to 62.90%. Compared to the literature in the same system, Jiang et al. [21] also studied the CFD simulation of microplastic separation with 1–75 μm particle diameters by using two-stage hydrocyclones and found the % recovery in the range of 45–72% with 10–20 μm particle diameters in the first hydrocyclone. The % recovery of PET particles, which have a higher density than PS, showed a higher % recovery because heavier particles have more inertia and are less influenced by the vortex. Increasing the liquid velocity inlet resulted in an increased % recovery due to the resulting increase in centrifugal force, which pushed the particles to the wall and moved them down to the underflow outlet. In terms of the split ratio (ratio of the volume of the underflow stream to the volume of the feed stream), each case had a split ratio of 0.56 ± 0.01%. The mean split ratio was depending on hydrocyclone design more than the operating conditions. However, when comparing the microplastic concentration, the results showed a small difference between 0.1% vol and 0.01% vol.

The ANOVA test was used to confirm the interaction between study parameters and microplastic separation in the hydrocyclone. Table 2 summarizes the results of the analysis of variance (ANOVA) test when the response variable was % microplastic recovery. The table displays the sums of square, degree of freedom, mean square, F-value, and P-value. A P-value below 0.05 was commonly considered statistically significant, resulting in the rejection of the null hypothesis. Conversely, if the P-value exceeded 0.05, it suggested that the deviation from the null hypothesis was not statistically significant, and hence, the null hypothesis was not rejected.

The main factor effects were analyzed, and it was found that the microplastic density (PS and PET) and inlet velocity had P-values lower than 0.05. Therefore, microplastic density and inlet velocity were considered significant factors in the % plastic recovery. When the density of plastics increased from 1,050 to 1,250 kg/m^3, plastics recovery increased from 64.48% to 77.31%, and the increase in inlet velocity resulted in an increase in plastic recovery from 68.68% to 73.47% (Fig. 4). The enhancement of microplastic density and inlet feed velocity facilitated effective separation in the hydrocyclone in a positive direction. For the system hydrodynamics, Fig. 5 shows the contour of the solid volume fraction within the hydrocyclone. The results indicated that the majority of the solid material was centrifuged towards the hydrocyclone wall. Furthermore, Fig. 6 depicts the response surface of plastic type and inlet velocity on microplastics separation (% plastic recovery). Highest separation of PET particles (density 1,250 kg/m^3) was subjected to a velocity of 10 m/s. This confirms that a high inlet velocity and high plastic density yield effective microplastic separation.

Table 2. The ANOVA test of microplastic recovery in hydrocyclone.

Parameter	Sum of squares	DF	Mean square	F-value	P-value
A	0.0311	1	0.0311	71.01	0.0004
B	0.0046	1	0.0046	10.5	0.0229
Residual	0.0022	5	0.0004		
Cor Total	0.0378	7			

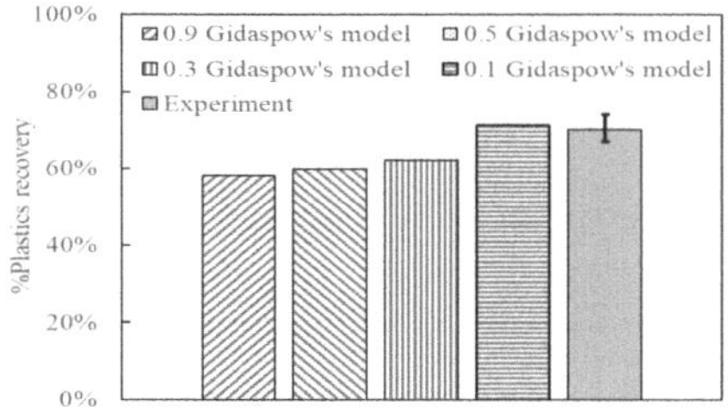

Fig. 3. Plastic recovery with difference drag coefficient modifications.

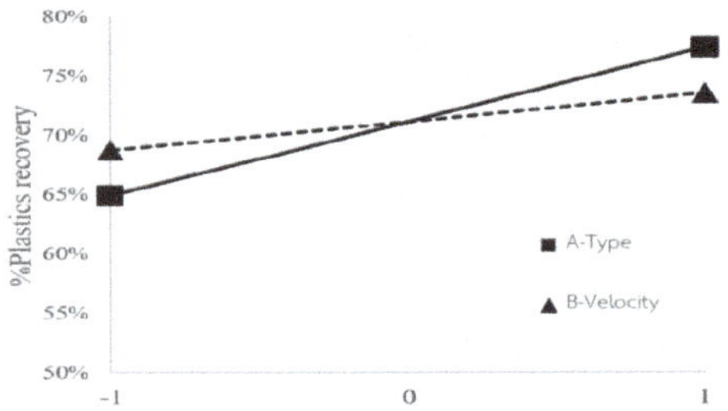

Fig. 4. Main effect plot of % plastic recovery in hydrocyclone.

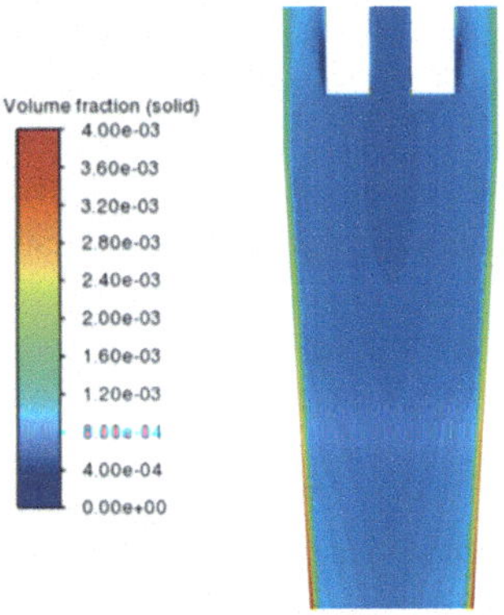

Fig. 5. Solid volume faction contour in hydrocyclone.

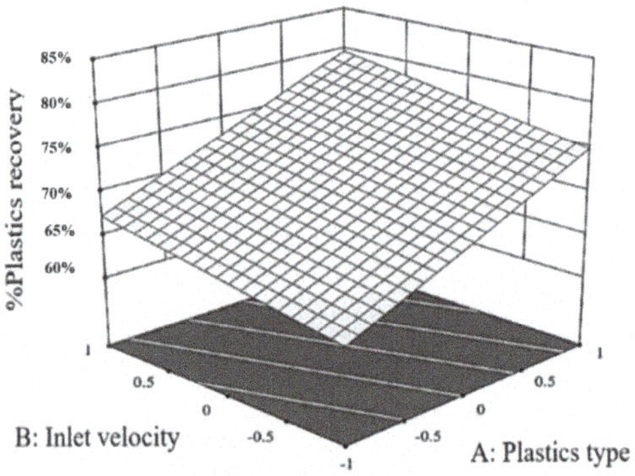

Fig. 6. Response surface on % plastic recovery in hydrocyclone.

5 Conclusion

The study aimed to optimize the number of computational cells and simulation time for the CFD model of a hydrocyclone for microplastic separation. The results indicated that 100,000 computational cells and 3 s of simulation time provided accurate and efficient simulation. The developed CFD model was successfully validated using experimental results. The hydrocyclone performance was analyzed with different types of microplastics, feed velocities, and plastic concentrations. It was observed that PET microplastics had a higher recovery than PS microplastics due to their higher density. The ANOVA test demonstrated that microplastic density and inlet feed velocity were significant factors that affected microplastic separation in the hydrocyclone. The study found that the PET particles with a velocity of 10 m/s, and concentration of 0.1%vol had the highest % recovery. Overall, the study provided valuable insights into optimizing the CFD model and understanding the factors that affected microplastic separation in the hydrocyclone.

Acknowledgment. The authors express their gratitude to the Thailand Research Fund and the National Research Council of Thailand (NRCT) for supporting this work through the Royal Golden Jubilee Ph.D. Program (N41A640353).

References

1. Frias, J.P.G.L., Nash, R.: Microplastics: finding a consensus on the definition. Mar. Pollut. Bull. **138**(November 2018), 145–147 (2019)
2. Koelmans, A.A., Mohamed Nor, N.H., Hermsen, E., Kooi, M., Mintenig, S.M., De France, J.: Microplastics in freshwaters and drinking water: critical review and assessment of data quality. Water Res. **155**, 410–422 (2019)
3. Chen, F., et al.: Impact of intensive mariculture activities on microplastic pollution in a typical semi-enclosed bay: Zhanjiang Bay. Mar. Pollut. Bull. **176**(March 2021), 113402 (2022)
4. Kuang, S.B., Chu, K.W., Yu, A.B., Vince, A.: Numerical study of liquid-gas-solid flow in classifying hydrocyclones: effect of feed solids concentration. Miner. Eng. **31**, 17–31 (2012)
5. Hildebrandt, L., Zimmermann, T., Primpke, S., Fischer, D., Gerdts, G., Pröfrock, D.: Comparison and uncertainty evaluation of two centrifugal separators for microplastic sampling. J. Hazard. Mater. **414**(December 2020) (2021)
6. Liu, L., et al.: Microplastics separation using stainless steel mini-hydrocyclones fabricated with additive manufacturing. Sci. Total. Environ. **840**(March), 156697 (2022)
7. Vakamalla, T.R., Koruprolu, V.B.R., Arugonda, R., Mangadoddy, N.: Development of novel hydrocyclone designs for improved fines classification using multiphase CFD model. Sep. Purif. Technol. **175**, 481–497 (2017)
8. He, L., Ji, L., Sun, X., Chen, S., Kuang, S.: Investigation of mini-hydrocyclone performance in removing small-size microplastics. Particuology **71**, 1–10 (2022)
9. Elsayed, K., Lacor, C.: Modeling, analysis and optimization of aircyclones using artificial neural network, response surface methodology and CFD simulation approaches. Powder Technol. **212**(1), 115–133 (2011)
10. Luo, H., Zhang, C., Sun, Z., Zhu, J.: Numerical investigation of hydrodynamics in liquid-solid circulating fluidized beds under different operating conditions. Adv. Powder Technol. **32**(4), 1047–1059 (2021)
11. Liu, W., et al.: A review of the removal of microplastics in global wastewater treatment plants: Characteristics and mechanisms. Environ. Int. **146**, 106277 (2021)
12. Li, C., Busquets, R., Moruzzi, R.B., Campos, L.C.: Preliminary study on low-density polystyrene microplastics bead removal from drinking water by coagulation-flocculation and sedimentation. J. Water Process. Eng. **44**(August), 102346 (2021)
13. Wenzel, M., et al.: Efficient and sustainable microplastics analysis for environmental samples using flotation for sample pre-treatment. Green Anal. Chem. **3**(October), 100044 (2022)

14. Sembiring, E., Fajar, M., Handajani, M.: Performance of rapid sand filter - single media to remove microplastics. Water Supply **21**(5), 2273–2284 (2021)
15. Bayo, J., López-Castellanos, J., Olmos, S.: Membrane bioreactor and rapid sand filtration for the removal of microplastics in an urban wastewater treatment plant. Mar. Pollut. Bull. **156**(April), 111211 (2020)
16. Rahaman, M.S., Choudhury, M.R., Ramamurthy, A.S., Mavinic, D.S., Ellis, N., Taghipour, F.: CFD modeling of liquid-solid fluidized beds of polydisperse struvite crystals. Int. J. Multiph. Flow **99**, 48–61 (2018)
17. Hamidian, A.H., Ozumchelouei, E.J., Feizi, F., Wu, C., Zhang, Y., Yang, M.: A review on the characteristics of microplastics in wastewater treatment plants: a source for toxic chemicals. J. Clean. Prod. **295**, 126480 (2021)
18. Magni, S., et al.: The fate of microplastics in an Italian wastewater treatment plant. Sci. Total. Environ. **652**, 602–610 (2019)
19. Eerkes-Medrano, D., Thompson, R.C., Aldridge, D.C.: Microplastics in freshwater systems: a review of the emerging threats, identification of knowledge gaps and prioritisation of research needs. Water Res. **75**, 63–82 (2015)
20. Adnan, M., Sun, J., Ahmad, N., Wei, J.J.: Comparative CFD modeling of a bubbling bed using a Eulerian-Eulerian two-fluid model (TFM) and a Eulerian-Lagrangian dense discrete phase model (DDPM). Powder Technol. **383**, 418–442 (2021)
21. Jiang, L., Liu, P., Zhang, Y., Yang, X., Wang, H., Gui, X.: Design boundary layer structure for improving the particle separation performance of a hydrocyclone. Powder Technol. **350**, 1–14 (2019)

Milling Parameters Identification Based on Known Surface Roughness

P. A. Senin[1], V. D. Gorodinov[2], and M. S. Kuts[1](✉)

[1] Department of Fundamentals of Machine Design, Bauman Moscow State Technical University, Moscow, Russian Federation
kuts@bmstu.ru
[2] Department of Plasma Power Plants, Bauman Moscow State Technical University, Moscow, Russian Federation
gorodinovvd@student.bmstu.ru

Abstract. The paper introduces a model for the identification of milling parameters, leveraging known surface roughness parameters. The accurate surface geometry was built by using the Z-buffer algorithm, which enables modeling according to specified cutting modes and tool geometries while also allows accounting the additional tool deflections caused by vibrations. For quantitative characterization of the obtained surfaces, a computational module that calculates roughness parameters in accordance with GOST ISO 4287-2014. GOST ISO 4287-2014 was developed. To address the inverse problem of milling parameter identification from known roughness, a neural network model based on a multilayer perceptron was trained using generated surfaces that incorporate vibration data. The development of the algorithms and neural networks was made using the Julia programming language and the Flux library. The results demonstrate an error margin of no more than 12% in mill diameter identification and no more than 5% in the identification of cutting modes. The developed neural network model, in conjunction with the surface generator based on the Z-buffer algorithm, offers a powerful tool for generating representative surfaces with specified roughness, that could be used in numerical simulations of contact interactions. Moreover, this model holds significant potential as a supportive tool in manufacturing engineering, particularly in the preproduction phase.

Keywords: Milling · Roughness · Surface modeling · Cutting parameters identification

1 Introduction

During design of machines and devices, it is crucial to consider the processes occurring at the contacts of the jointed parts. It is well established that the geometry of parts surfaces is not ideal, comprising a multitude of macro- and micro irregularities of the surface geometry. However, the contact layer of joints exhibits nonlinear elastic properties that differ from the theoretical elastic properties of the part [1, 2].

© The Author(s), under exclusive license to Springer Nature Singapore Pte Ltd. 2025
J. Mo and P. Chutima (Eds.): ICMSC 2024, LNME, pp. 97–105, 2025.
https://doi.org/10.1007/978-981-96-5380-5_10

Contact stiffness also plays a significant role in bolted joints, widely used in mechanical engineering. As it was shown in [3], the presence of irregularities in the contact layer results in an increase in its pliability, causing blurring the pressure in the bolt vicinity. Furthermore, in [4], the damping and dynamic stiffness of bolted joints were investigated, and in [5], it was shown that the external load on the joint decreases with increasing screw tightening force. This is a result of the nonlinear relation between contact displacements and contact pressure. The reason of this nonlinearity is the increase in actual contact area during the convergence of the contact surfaces, which usually have complex geometry due to the manufacturing process. The impact of surface geometry on contact stiffness was revealed in [6]. In [7], the dependences of the relative actual contact area of flat contacting surfaces of steel parts on the nominal pressure under dynamic contact loading were investigated, and in [8] a method for calculating the convergence and contact stiffness of parts of double curvature of close hardness was proposed. Thus, it is essential to have an accurate and realistic representation of the contact surface geometry when dealing with the modeling of contact interaction.

Among the approaches to surface modelling following machining, two main groups can be distinguished: statistical, which consider the surface as a random process with a certain distribution law, and simulation, which reproduce the profile of the real surface.

In the first case, the calculation of the contact layer stiffness is performed by means of simplified computations in which irregularities are represented by a set of geometric bodies (spheres, ellipsoids, prisms or pyramids) distributed according to a certain law [9]. This class of models also includes models based on probabilistic modelling of surfaces with subsequent spline interpolation [10, 11].

The second method has an indisputable advantage in terms of accuracy compared to the first ones, as it takes into account the real shape of micro-irregularities. Among the methods used for modelling the real profile, two approaches are the most prevalent: 1) the one based on usage of constructive solid geometry (CSG) [12, 13], and 2) the approach based on utilization of Z-buffer [14]. The disadvantage of this class of methods is that when modelling, it is necessary to specify cutting parameters (cutting speed, feed rate, cutting depth and geometric parameters of the mill), that determine the microgeometry of the modelled surface. Furthermore, it is important to consider the vibrations that occur during the cutting process, as they change the actual trajectory of the tool movement and also have a significant impact on the microgeometry.

This paper considers the problem of identifying the tool geometry and cutting parameters during side milling. A model based on a neural network has been developed, which enables the selection of milling parameters using known roughness parameters. This is necessary to obtain a three-dimensional surface with specified roughness characteristics. The surface thus obtained can be further used in the calculation of the contact layer stiffness by the finite or boundary element method. The open-source library Flux was used to build the model. The model was trained based on data obtained by direct modelling of the surface using the Z-buffer algorithm with different variants of initial values of milling parameters and tool geometry.

2 Methodology

2.1 Geometry Generation

Modelling the Movement of the Milling Cutter. The initial parameters for surface modelling were the geometry of the cutting edge and the cutting modes during milling. The cutter geometry was specified by the following parameters: cutter diameter, cutting edge length, helical line pitch.

According to these data, the helical line of the cutting edge of the cutter was discretised and the coordinates of the obtained points were recorded in an array. In homogeneous coordinates, this array of N points is defined as:

$$P_{cl} = \begin{bmatrix} x_1 \; x_2 \; \dots \; x_N \\ y_1 \; y_2 \; \dots \; y_N \\ z_1 \; z_2 \; \dots \; z_N \\ 1 \;\; 1 \; \dots \;\; 1 \end{bmatrix} \tag{1}$$

The array was sequentially multiplied by the rotation matrix in order to model the rotation of the cutting tool.

$$M_{ri} = \begin{bmatrix} \cos(\varphi_i) & -\sin(\varphi_i) & 0 & 0 \\ \sin(\varphi_i) & \cos(\varphi_i) & 0 & 0 \\ 0 & 0 & 1 & 0 \\ 0 & 0 & 0 & 1 \end{bmatrix} \tag{2}$$

where $\varphi_i = \omega t_i$ is the angle of rotation of the cutter around its axis at the i-th moment of time $t_i = i\Delta t$.

The linear movement of the cutter was obtained by multiplying the rotation matrices obtained previously by the translation matrix:

$$M_{ti} = \begin{bmatrix} 1 & 0 & 0 & s_{xi} \\ 0 & 1 & 0 & s_{yi} \\ 0 & 0 & 1 & s_{zi} \\ 0 & 0 & 0 & 1 \end{bmatrix} \tag{3}$$

where s_{xi}, s_{yi}, s_{zi} - the value of linear movement of the cutter relative to the initial position at the i-th moment of time along the corresponding coordinates.

Consequently, a three-dimensional matrix was generated, representing the successive positions in space of the cutter cutting edge during the machining time. The surface was formed by combining the neighbouring 8 points into polygons, each of which was described by a second-degree polynomial [14]. To obtain the milling surface, it is necessary to filter out the lower part of this surface excluding self-intersection. For this purpose, the Z-buffer algorithm was employed.

Z-Buffer. The fundamental principle underlying this method is as follows. The direction of projection to the surface for which a model must be created using the Z-buffer method is selected, and the plane of the future Z-array is then drawn perpendicular to this direction. On the selected plane, the position of the rectangular area limiting the projection of the surface on the Z-array plane is determined. One of the corners of the obtained rectangle is chosen as the origin point of the Z-array with coordinates (x_0, y_0). Inside the rectangular area, a regular grid of rectangular elements is constructed. For the center of all grid cells, the Z coordinate of the corresponding surface point is determined, which is the distance from the Z-array plane along the its normal to the intersection with the approximated surface. This value is stored in a two-dimensional array. It is assumed that the Z-level (depth level) within a single grid cell is constant. When cutting the material, the position of the grid cells does not change, only their Z-coordinate changes. An important feature is the constancy of the grid step, which is necessary for the construction of fast algorithms for determining the thickness of the cut layer and building new surfaces. Z_buffer algorithm has been previously used for modelling the milling process in [15–17]. A detailed description of the algorithm can be found in [14, 18]. The implementation of this algorithm was conducted in the Julia programming language [19].

2.2 Calculation of Roughness Parameters

A set of parameters should be employed to describe the surface profile. This set should comprise a minimum number of values that collectively provide the most comprehensive description of the surface profile and do not duplicate each other. In the course of the work, five main parameters (as defined in GOST R ISO 4287–2014) were selected for analysis: the arithmetic mean deviation of the profile *Ra*, the asymmetry of the estimated profile *Rsk*, the average width of profile elements *Rsm*, the maximum height of profile protrusion *Rp*, and the maximum depth of depression *Rv*. Additionally, an algorithm for calculating these parameters for surfaces was implemented in the Julia programming language.

2.3 Consideration of Vibrations

It is inevitable that vibrations will occur during the milling process, which will contribute to the roughness of the machined surface.

The cause of vibrations during the cutting process is the elastic deformation of the material in the cutting zone, as well as the fluctuation of cutting force caused by the periodic entry and exit of the cutting edge of the tool into the cutting zone. Consequently, in order to construct a functional surface identification model, it is necessary to take the vibrations into account. At the same time, it is worth mentioning that the development of a vibration model is beyond the scope of current work. Thus, in the current work, the simplified approach to model vibration was used just to reproduce the vibration observed during the real milling process.

In order to obtain the reference data, an experiment was conducted on the machining of workpieces on a Rabbit milling machine with an Iscar end mill of 8 mm diameter with 3 cutting edges. During the milling process at the selected modes, the amplitude spectrum of the vibrations generated during the cutting process was measured using a DeltaTron type 4397 accelerometer from Brüel & Kjær, and the surface profile was measured using a TR220 profilometer.

The Fast Fourier Transform (FFT) was utilized to obtain the vibration spectrum, which could then be conveniently included in the surface modelling using the Z-buffer method. Having obtained the amplitude, frequency and phase of each harmonic, it is possible to represent them as a sinusoidal signal. The sum of the sinusoidal displacements at each moment of time is written into a displacement matrix. To obtain displacements from the vibration acceleration signal, the signal is integrated twice and the data from the experiment is expressed as displacements along the axis. Based on preliminary modelling, it was determined that the six largest amplitudes of the spectrum are sufficient to describe the vibrations observed during the experiment. By including the vibrations into the computations, the modelled surface becomes similar to the real one. Figure 1 presents a comparison between the real surface and the generated one, obtained without taking into account the vibrations (Fig. 1-a) and with additional mill fluctuations caused by the vibrations (Fig. 2-b). In the vibrationless version, the amplitude characteristics were derived from the experiment and the phase components was manually tuned to achieve better fitting. It is important to note that the contribution of vibrations in this case is significantly greater than the contribution of cutting kinematics and tool geometry. This is explained by the fact that at high-speed cutting used in the experiment, the tool feed is many times less than the cutting speed.

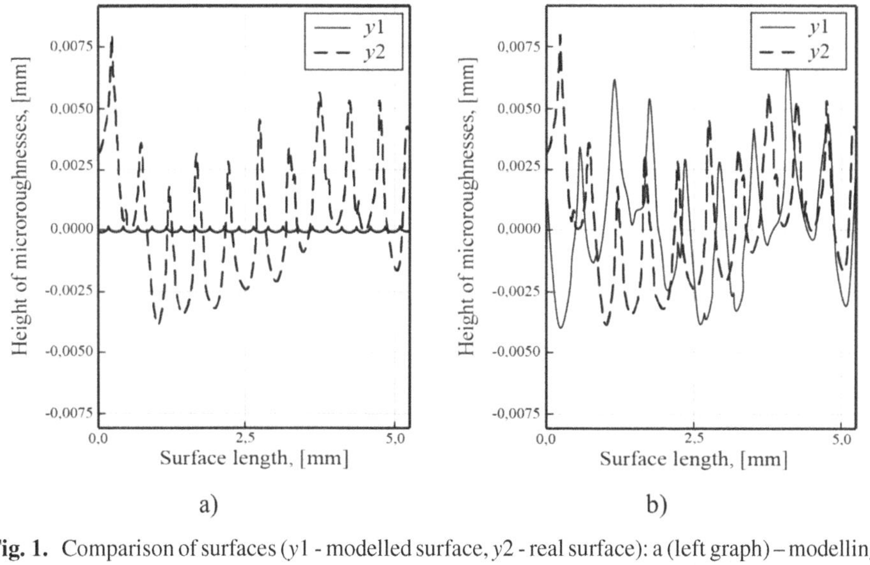

Fig. 1. Comparison of surfaces ($y1$ - modelled surface, $y2$ - real surface): a (left graph) – modelling result without vibration; b (right graph) – the result of the modelling accounting vibrations.

It is also worth noting that although the specific realization of the surfaces may not coincide (Fig. 1-b), in this particular case we are interested in how close the surfaces are from a statistical point of view. In this case, the most complex parameter is the curve of the reference surface (Fig. 2), which shows that the real and generated surfaces have rather close topography.

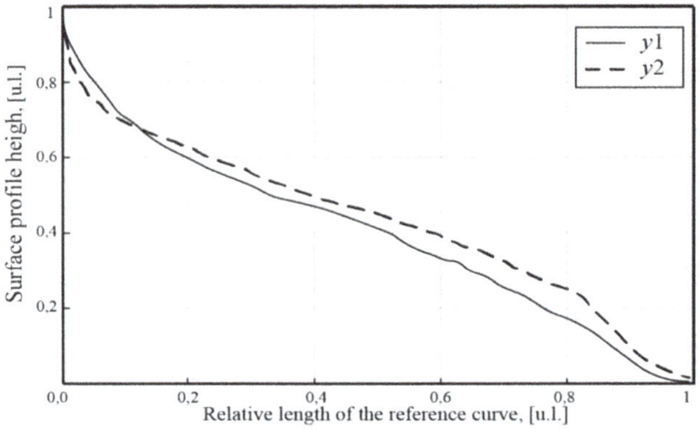

Fig. 2. Comparison of surfaces by reference curves (*y*1 - modelled surface, *y*2 - real surface).

2.4 Neural Network Model

To obtain the optimal milling parameters according to the given surface roughness parameters, a neural network of direct propagation (multilayer perceptron) was used. The model is a multilayer structure approximating the function $Y = F(R)$, where Y - milling parameters, R - roughness parameters. Contains 4 hidden layers: the first one consisting of 10 neurons with sigmoid activation function, the second one consisting of 20 neurons with hyperbolic tangent activation function, the third one containing 30 neurons with ReLu activation function, and the fourth 10-neuron layer with sigmoid activation function. The architecture of the model is shown in Fig. 2.

Using the Z-buffer algorithm, 980 surfaces were modelled for training the model and 20 for testing its performance, with the input parameters determined randomly in the ranges: mill radius $r = 5 \ldots 32$ mm; cutting depth $t = 0.5 \ldots 1.5$ mm; feed rate $v_m = 1000 \ldots 5000$ m/min; cutting speed $n = 2500 \ldots 22000$ rpm. For each modelled surface 5 roughness parameters were calculated using the previously developed computational module.

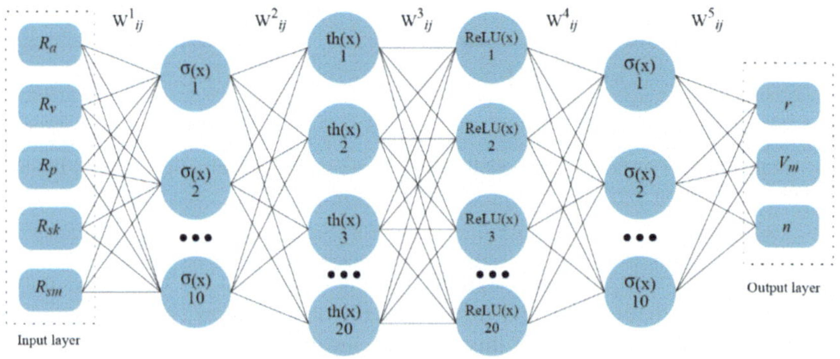

Fig. 3. Model architecture.

The Flux library for the Julia programming language was employed to implement the algorithm in question. The Adam optimiser, which incorporates adaptive moment estimation, was selected as the neural network optimiser.

3 Results and Discussion

3.1 Training Result of the Model

Although the cutting depth affects the milling process in general, in the case of decoupled vibration model from surface modeling, it can be excluded from input parameters as it does not affect final surface profile at same vibration level. Therefore, the output layer can be reduced to 3 parameters. To model the effect of vibration, the signal obtained from computational experiment on simulation of milling process is used, based on model described in [20].

The analysis of the model results concerning virtual surfaces showed good convergence of the predicted values for the milling cutter radius r, feed rate v_m and rotational speed n with those given for surface modelling by the Z-buffer method initially. The obtained results after normalization are shown in Fig. 3.

For each value, a relative error in percent was calculated. The best convergence is observed for the mill radius. The error did not exceed 15%. The feed ratio and cutting speed have larger discrepancies, but they can still be considered to be within the normal range, which can be explained by their less significant influence on the final roughness parameters relatively to the influence of the mill radius. Thus, the predicted feed rate exceeded the 15% error only in 2 out of 20 cases, and the rotational speed - in 6 out of 20 cases.

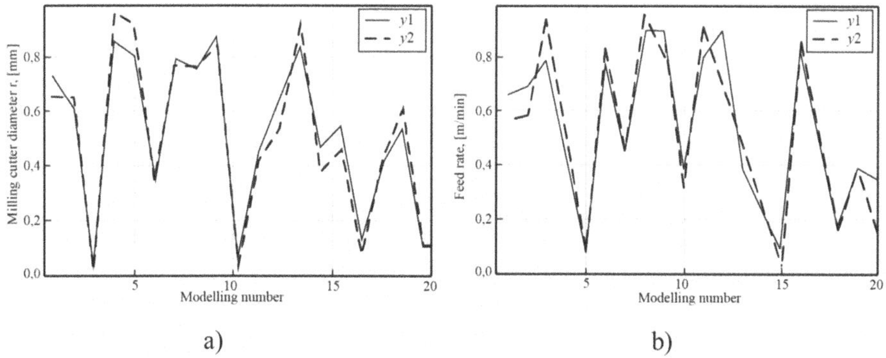

Fig. 4. Comparison of identified ($y1$) mill radius values and tool feed rates with the actual ones ($y2$): a) - the milling cutter diameter r; b) - feed rate v_m.

3.2 Verification of the Model Performance

To check the performance of the model, we built a surface profile using the generated data and compared it to the initial surface profile. For this purpose, a following test data

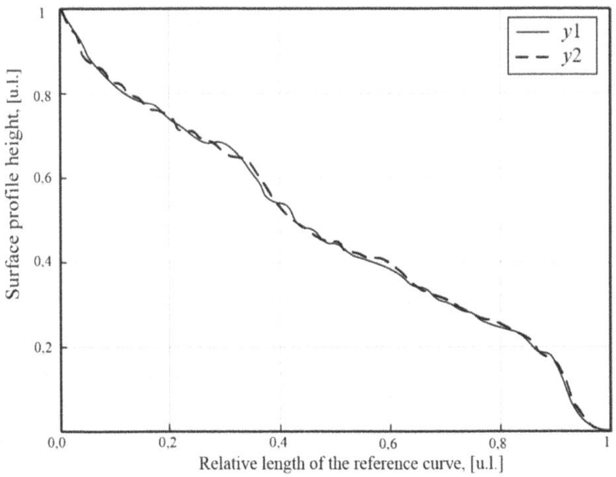

Fig. 5. Comparison of reference curves of surface profiles for the initial (*y1*) and reconstructed (y2) surfaces.

set was used: mill radius $r = 30{,}0$ mm; feed rate $V_m = 1247{,}0\frac{\text{mm}}{\text{min}}$; rotational speed $n = 14079{,}0$ rpm. The surface profile was obtained using these data.

As a result, the following parameters were obtained: cutter radius $r = 26{,}98775$ mm; feed rate $V_m = 1193,857\frac{\text{mm}}{\text{min}}$; rotational speed $= 13767{,}07$ rpm. After that, a new surface profile was created using the algorithm described in Sect. 2.1. Figure 4 reveals the comparison of the obtained and initial profile by reference curves.

4 Conclusions

This paper presents a model that identifies the tool geometry and the milling parameters used to obtain it, given roughness parameters. As a result of the study the following results were obtained: considering the virtual surface obtained from modelling with known milling parameters, the model shows good convergence of predicted cutting parameters. For the mill radius the average error was 10.4%, for the feed rate – 4.3%, for the rotational speed – 2.2%. The average error of the initial and obtained roughness parameters was 1,7%. These values are in the vicinity of 10%, so the model can be considered as workable. The difference in the error of identification of the cutter radius and cutting modes indicates that for high-speed machining the mill radius and the vibrations occurring during machining have the greatest influence on the surface roughness.

The RMS error of the model was less than 0,1 μm. Comparison of the roughness parameters of these surfaces showed that the average error was 1,7%.

The obtained model can be used to identify the contact stiffness of mechanical joints by modelling the contact interaction between the surfaces of mating parts (Fig. 5).

References

1. Suslov, A.G.: Razvitiye ucheniya o kontaktnoy zhestkosti i inzhenerii poverkhnosti detaley mashin. Vestnik Bryanskogo gosudarstvennogo tekhnicheskogo universiteta **11**(72), 12–17 (2018)
2. Andrienko, L.A., Zemskov, A.A., Kuts, M.S.: Vibrational diagnostic indicators of assembly defects in transmissions. Russ. Eng. Res. **43**(4), 404–409 (2023)
3. Kuts, M.S.: Pressure distribution in the area around a tightened single bolt joint. Izvestiya vysshikh uchebnykh zavedeniy. Mashinostroyeniye **1**(706), 3–11 (2019)
4. Andrienko, L., Kuts, M., & Chirkin, A.: Dynamic properties of filled bolted joints. In: Dynamics of Technical Systems 2021. DTS, vol. 2507, no. 1. AIP Publishing (2021)
5. Ivanov, A.S., Kuts, M.S., Murkin, S.V.: Relation between the tightening force and the external load on a screw in a threaded joint. Russ. Eng. Res. **42**(4), 316–322 (2022)
6. Tikhomirov, V.P., Izmerov, M.A.: Vliyaniye volnistosti i sherokhovatosti poverkhnosti na normalnuyu kontaktnuyu zhestkost ploskogo styka. Transportnoye mashinostroyeniye **12**(85), 4–12 (2019)
7. Matlin, M.M., Kazankin, V.A., Kazankina, E.N., Mozgunova, A.I.: Contact stiffness of mating surfaces of parts under dynamic loading. Vestnik mashinostroyeniya (6), 57–60 (2020)
8. Matlin, M.M., Mozgunova, A. I., Kazankina, E.N., Kazankin, V.A.: Contact stiffness of double curvature parts with close hardness. Izvestiya Volgogradskogo gosudarstvennogo tekhnicheskogo universiteta (2), 83–88 (2016)
9. Voronov, S.A., Kiselev, I.A., Voronova, I.S.: A stochastic model of plane grinding dynamic for the texture formation analysis. Vibroeng. Procedia **38**, 185–192 (2021)
10. Ogar, P.M., Gorokhov, D.B.: Contact interaction of rough surfaces: Fractal Approach. Sistemy. Metody. Tekhnologii (2). 30–38 (2010)
11. Wei, C., Zhu, H., Lang, S.: A modified complete normal contact stiffness model of a fractal surface considering contact friction. Fractals **28**(05) (2020)
12. Dreval, A.E., Vasiliev, S.G., Vinogradov, D.V.: Package for modeling tool and structural materials wear. In: Modeling in Engineering 2020, vol. 2383, no. 1. AIP Publishing (2022)
13. Potapova, M.S., Vinogradov, D.V.: Kompyuternoye modelirovaniye relyefa poverkhnosti. obrabotannoy frezoy s krivolineynoy rezhushchey kromkoy. Mashinostroyeniye i kompyuternyye tekhnologii (6), 42–55(2015)
14. Kuts, M.S., Chirkin, A.V.: An approach for modelling of rough surface after milling using quadratic serendipity elements and Z-buffer. JARiTS **33** (2022)
15. Kuts, V., Kiselev, I., Voronov, S.: Ensuring the chatter stability of milling by optimizing the workpiece stock. Vibroeng. Procedia **38**, 179–184 (2021)
16. Roth, D., Ismail, F., Bedi, S.: Mechanistic modelling of the milling process using an adaptive depth buffer. Comput.-Aided Design **35**(14), 1287–1303 (2003)
17. Saito, T., Takahashi, T.: NC machining with G-buffer method. ACM Siggraph Comput. Graph. **25**(4), 207–216 (1991)
18. Greene, N., Kass, M., Miller, G.: Hierarchical Z-buffer visibility. In: 20th Annual Conference on Computer Graphics and Interactive Techniques, pp. 231–238 (1993)
19. Bezanson, J., Edelman, A., Karpinski, S., Shah, V.B.: Julia: a fresh approach to numerical computing. SIAM Rev. **59**(1), 65–98 (2017)
20. Zavarzin, D.A., Kiselev, I.A., Tsiganov D.L.: Modelirovaniye protsessa ploskogo frezerovaniya s uchetom zavisimosti dinamicheskikh kharakteristik stanka. Mashiny i ustanovki: proyektirovaniye. razrabotka i ekspluatatsiya (4), 53–68 (2016)

An Experimental Study of the Vortical Structures in Supersonic Steam Jet Impinged on to the Wall

Afrasyab Khan[1](✉), Khairuddin Sanaullah[2], Pavel Alexandrovich Taranenko[3], Vladimir Vladimirovich Sinitsin[3], David Bassir[4,5](✉), Ajmal Shah[6], Zafar Ahmed[7], and Andrew Ragai Henry Rigit[8]

[1] University of Dundee, Dundee, UK
akhan004@dundee.ac.uk
[2] University of KwaZulu-Natal (UKZN), Glenwood, Durban, South Africa
[3] South Ural State University (SUSU), Chelyabinsk, Russian Federation
[4] Smart Structural Health Monitoring and Control Laboratory, DGUT-CNAM, Dongguan University of Technology, Dongguan, China
david.bassir@utbm.fr
[5] ENS -Paris-Saclay University, Centre Borelli, UMR CNRS 9010, 91190 Gif-sur-Yvette, France
[6] Pakistan Institute of Engineering and Applied Sciences (PIEAS), Islamabad, Pakistan
[7] Northumbria University, Newcastle Upon Tyne, UK
[8] University Malaysia Sarawak (Unimas), Kota Samarahan, Malaysia

Abstract. This article addresses the issue of supersonic steam jet interaction with the wall. Steam has been impinged in a wall using a nozzle at different pressure values ranging from 1.5–3.0 bars. The whole fluid domain is comprised of regions, that are the body of the jet and the fluid region near the wall. Within this region, the transformation of primary vortices into secondary vortices occurs and these vortices expel outward in the radial direction after impingement. Measurements have shown the spreading of the normalized density of the vortical structures along the axial length within the region near the wall. However, the probability density function analysis has indicated that the vortical structures become denser within the impingement region. In addition to the steam interaction with wall, along the radial direction, weak traces of their existence in the jet region may be attributed to the vorticity diffusion within the jet region.

Keywords: Supersonic · Steam; Jet · Vorticity; Swirl · Wall

Nomenclature

Symbol	Description
L	Axial length (m)
r	Radial length (m)
D	Nozzle diameter (m)
t	tTime (s)
u	uAxial velocity (m/s)
B	Sum of external forces

© The Author(s), under exclusive license to Springer Nature Singapore Pte Ltd. 2025
J. Mo and P. Chutima (Eds.): ICMSC 2024, LNME, pp. 106–122, 2025.
https://doi.org/10.1007/978-981-96-5380-5_11

Greek symbols

ω Vorticity
ρ Fluid density (kg/m^3)
Viscous stress tensor
Optimal in-homogeneous proper orthogonal decomposition (POD) basis function for the axial component of the velocity
Ω Swirling strength

Subscripts

x Along axial axis
r Along radial axis
θ Ensemble averaged

1 Introduction

When a jet, either consisting of the same fluid in its surroundings or a different one, impinges onto a wall and spreads in the surroundings along the radial direction to the jet, the characteristics of such a fluid domain comprised of the characteristics of the free jet and fluid that exists in the region near the wall [1]. The jet will face retardation due to the existence of the wall and the fluid region. Thus, structures comprising vortices within such fluid domains are generated, which resemble closely with the structures that appear at the boundary of the interacting fluids.

Furthermore, it has been observed that structures within the vicinity of the central jet or across the jet stream have close resemblance with those appearing at the periphery and surroundings of such vortices. These have been termed as primary and secondary vortices. The fluid present within the jet repels from the central region, whereas the surrounding fluid is entrained by the jet. Their matched shapes and slight variation related to the boundary conditions [1]. With a huge body of research findings in this regard which included studies involving air traveling within water and smoke was used as tracer particles [2]. In current study, the impingement of jet onto a wall has been studied. Steam jet interactions with the structures have myriad applications that range from steam interaction with turbine blades, textile products manufacturing, steam usage in the metal processing industry and in the paper industry, and a lot more can be narrated in this regard. Yet all these application areas enlighten the steam interaction with some solid body/wall, however, for further attention on this topic till date, only few studies have narrated the properties of the domain, focusing interaction between them with the wall [3–6]. The vortical structures are formed when jet impinges onto wall at 90°, and such study requires detailed attention from the scientific community and current study addresses this issue. Jet impingement onto a wall was discussed in many studies, where the jet was generated using nozzle. On research grounds, such efforts the jet is considered as the main body of the fluid, which was stretched over 4–5 D (where D is the nozzle exit diameter) [7–9]. Other significant parameters being involved in such supersonic steam profile, include the turbulence level, the distance between the wall and the nozzle,

Reynold's number, and the shape of the nozzle along with the configuration of the jet [10–11]. The jet emerges from the nozzle. It can be characterized by the nozzle design, as even at a very lower pressure gradient it can produce the supersonic velocity profiles at its exit, as mentioned in our previous studies [12–16]. Following along the axial and radial directions (axial length L = 0 → L = ∞) & (r = 0 → r = r) respectively, the transformation of initially supersonic steam jet takes place from non-varying flow along the axial axis, along with the shearing layer at the exit of the nozzle, to a completely developed flow as far as we observe away from the exit. This type of phenomena includes such conditions where the jet expands along the radial direction after being interacted with the wall. With the inherent complexities which are inhibited by the regions inside and around the jet [8, 17–19] it's a complex task to simulate all these regions all at once with a single model for the turbulent flows [19]. The occurrence of the vortices and the interaction between them have a huge impact on the heat, momentum and mass transfer across the fluid domain and their interaction has been analyzed using Eq. 1 (Vorticity equation).

$$\frac{D\omega}{Dt} = \frac{\partial \omega}{\partial t} + (u.\nabla)\omega$$
$$= (\omega.\nabla)u - \omega(\nabla.u) + \frac{1}{\rho^2}\nabla\rho \times \nabla p + \nabla \times \left(\frac{\nabla.\mathrm{T}}{\rho}\right) + \nabla \times \left(\frac{B}{\rho}\right) \quad (1)$$

where ω is the vorticity, t is time (s), u is the axial velocity (m/s), ρ is the fluid density (kg/m^3), T is the viscous stress tensor and B is the sum of the external body forces. By using the experimental setup described in the latter section, the vortical distributions, the area under their influence as well as their swirling strengths are estimated. The details of the experimental setup and the outcomes extracted from the data are described in detail in the following sections.

2 Experimental Setup

The experimental setup is comprised of a nozzle and a rectangular Perspex box. The nozzle geometry is shown in Fig. 1. The capability of the nozzle with these dimensions has been confirmed to generate the supersonic jet flow in our previous studies [12–15, 20]. The CFD study involves application of steady state model in the Ansys Fluent. The steam has been simulated by using the Direct Contact Condensation (DCC) Model [21]. The physical flow details of the DCC model including the key equations and the operating parameters are followed. In Direct Contact Condensation (DCC) model, whereas the phases involved has distinct interpretation as continuous medium.

With the specified equations for each phase, that are related to momentum, continuity energy with the interaction between phases is represented with coupling has been achieved by the usage of the pressure coefficient and the interphase exchange coefficients. For computing the drag we use the symmetric model and for the turbulence we have use the realizable model between phases.

The equation for the continuity is given as,

$$\frac{\partial}{\partial t}(\alpha_q\rho_q) + \nabla.(\alpha_q\rho_q v_q) = \sum_{p=1}^{n}(\dot{m}_{pq} - \dot{m}_{qp}) + S_q \quad (2)$$

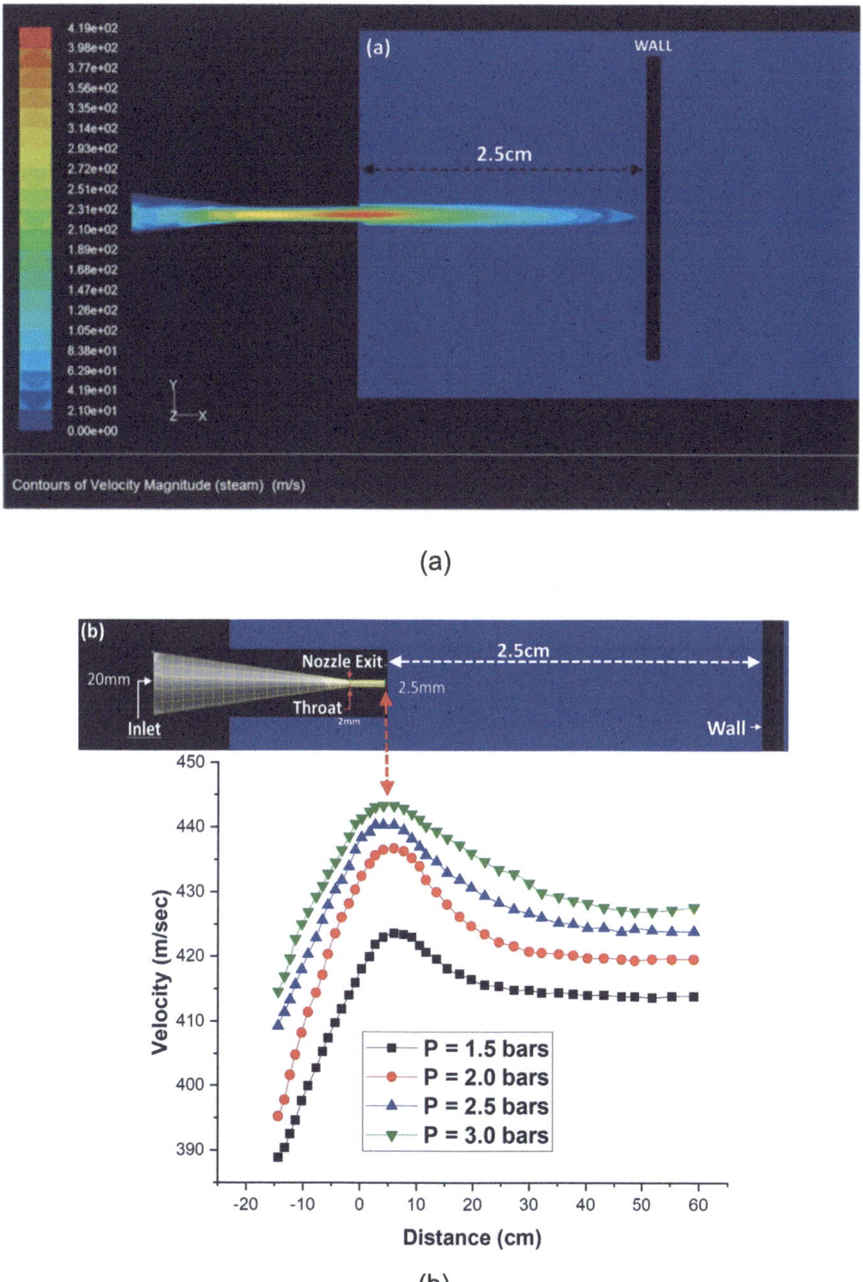

Fig. 1. (**a**) CFD Study based velocity contours [12–15, 20] (**b**) Supersonic Nozzle Design with velocity profiles.

whereas, v_q is the qth phase velocity, and the rate of mass transfer between two phases is given by $\dot{m}_{pq}$. Is Whereas the source term is given by S_q. The momentum transfer has been shown by the following equation;

$$\begin{aligned}
&\frac{\partial}{\partial t}(\alpha_q \rho_q v_q) + \nabla.(\alpha_q \rho_q v_q v_q) \\
&= -\alpha_q \nabla P + \nabla.\tau_q + \alpha_q \rho_q g \\
&+ \sum_{p=1}^{n} (R_{pq} + \dot{m}_{pq} v_{pq} - \dot{m}_{qp} v_{qp}) \\
&+ (F_q + F_{lift,q} + F_{vm.q})
\end{aligned} \tag{3}$$

The velocities between the phases is given by v_{qp} and v_{pq}. The body force is given by F_q and the stress tensor is mentioned by τ_q for the respective phase, and the lift and virtual mass forces are given by $F_{lift,q}$ and $F_{vm.q}$ respectively. The pressure is given by P and the interaction between the phases is given by R_{pq}. The equation used to govern the transfer of energy is given by;

$$\begin{aligned}
&\frac{\partial}{\partial t}(\alpha_q \rho_q h_q) + \nabla.(\alpha_q \rho_q u_q h_q) \\
&= -\alpha_q \frac{\partial P_q}{\partial t} + \tau_q : \nabla u_q - \nabla q_q + S_q \\
&+ \sum_{p=1}^{n} (Q_{pq} + \dot{m}_{pq} h_{pq} - \dot{m}_{qp} h_{qp})
\end{aligned} \tag{4}$$

whereas the source term, specific enthalpy and the heat flux are given by the terms S_q, h_q, and q_q. The heat exchange intensity is mentioned by Q_{pq} and the exchange of enthalpy is mentioned by h_{qp}. The details regarding the operating boundary conditions used in the simulation study are given in the Table 1 below.

With a pressure ranging from 1.5–3.0 bars, steam has been injected inside the water via a nozzle that is capable of producing the supersonic steam jet. With the dimensions of LHW = 4 × 2 × 3, the water has been accumulated using the Perspex box (Fig. 2). The nozzle is inserted into the box through a pipe insulated with heavy Teflon taping to preserve the quality of the injected steam. As the nozzle is at a distance of 2.5 cm from the wall, for the measurement of velocities we have used the particle image velocimetry (PIV) system. However, due to the physical limitations, just half of the whole region could be scanned using the PIV technique. This has been caused by the body of the nozzle hindering the laser light in the downward region of the nozzle. As the PIV system has a CCD diode camera that has the resolution of 7.4 μm^2 ~ 2048 × 2048 pixels.

With a power of 200 mJ/pulse and the pulse frequency of ~15Hz, the laser has been used with the dedicated software for the processing of data and the National electronic DAQ system. We have used as tracer particles the fluorescent 1.0g/cc microspheres, with the laser sheet being inclined vertically and the camera is being installed in the horizontal direction, the given setup has been used for the acquisition of experimental data. With a total of six thousand frames being acquired in a time range of 10 min along a single operating condition.

Table 1. Boundary conditions used for simulation study

Phase	Density (kg/m3)	Dynamic Viscosity ((Pa s, N s/m2) x 10^{-3})	Kinematic Viscosity ((m2/s) x 10^{-6})	Young's modulus (GPa)	Poisson's ratio	pressure (bars)	Temperature (^{0}C)	Ultimate Tensile Strength (MPa)
Water	Constant value taken from	0.798-	0.801-	---	----	Outlet pressure= 1 bar	25^{0}C	----
Steam	Constant value taken from Fluentdata base	Fluent data base	Fluent data base	----	----	Nozzle Inlet = 1-3 bars	100-134^{0}C [22]	----

Model	**Solver:** Pressure based **Formulation:** Implicit	
Multiphase model	**Scheme:** Euler 2 Phase **Viscous Model:** k-epsilon 2 equations **k-epsilon:** Realizable	**Near wall treatment:** Standard wall function **k-epsilon multiphase model:** Mixture
Drag	**Steam-water:** Symmetric **Heat:** Ranz-Marshall **Mass transfer:** Will be determined by **User Defined Function (DCC model)** **Under relaxation factor:** 0.1 **Pressure-velocity coupling:** Phase coupled SIMPLE **Discretization:** First order upwind	
Multigrid control	**Cycle type:** F- Cycle **Termination restriction:** 0.1 **AGM method stabilization method:** Aggregative-BCGSTAB **Convergence criteria:** 1 e -5	
Direction of pressure specification method: Normal to the boundary	Turbulence specification method: k-epsilon	

The main focus area of the CCD camera is the most prominent region where the steam interacted with the wall, as shown in Fig. 2. The uneven points have been removed with the help of the anomaly detection. The laser pulse magnification and separation method

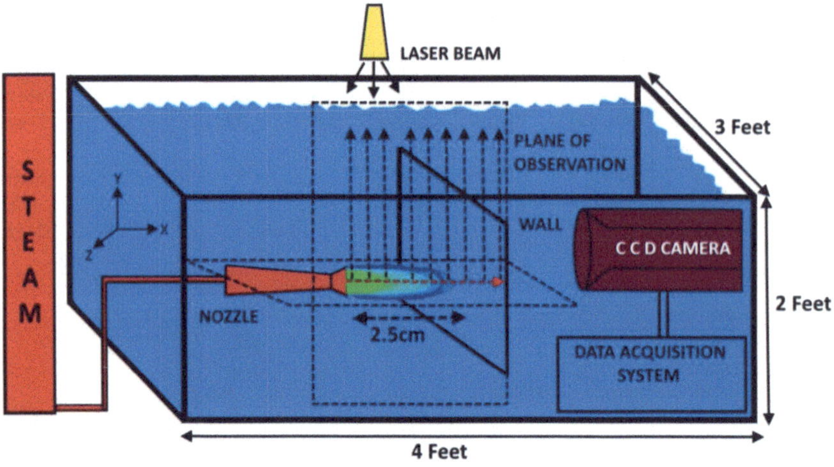

Fig. 2. A schematic of an experimental setup.

has been used for the calibration of the PIV system. Laser sheet has been focused on the photodetector, for the indication of the estimated separation, thus the distance has been calculated for the points of emergence and incidence of light beam on the photodetector. Thus, by taking the ratio of the known grid spacing and the know grid that dimensions we already know. Thus, the optical distortions have been used for the measurement of the errors in the magnification which in the current case remains in the range of 0.1–0.3%.

2.1 Mathematical Formulations

For the last two decades, the amount of information gathered using experimental and theoretical methods for the understanding of turbulence has grown appreciably. These include the common methods (e.g. hot wire method) used for the study of turbulence. Whereas on experimental grounds the PIV, and holography and on theoretical grounds Direct Numerical Simulation (DNS) and Large Eddy Simulation (LES) are the preferred methods. These methods discern well the dynamics, kinematics and scale resolution to extract the information related to the fluid domain. Reynold's stress decomposition into its constituents i.e. the mean and fluctuating parts provide enough insight into the flow domain.

Yet these methods may prove successful to deliver correct components of Reynold's stress, however, they have been proved not ideal to provide a clear understanding of the inhibiting flow process. The best method for drawing the complete picture of the process depends on the nature of the inhibited process and the goal to be achieved. In terms of statistical inter-relationship, Reynold's decomposition method may prove to be the best choice. In view of the understanding of the structures and elements of the flow, the visualization methods can provide a deeper insight into the flow structures. However, the interaction between the smaller eddies and their interaction with the large flow structures can be exhibited through the detection of the smaller eddies. Also, the visualization of the streak lines across a fluid domain has been conducted. Even having our weak confidence on such techniques to offer accurate flow picture, however, such information has been obtained here to exhibit characterization of physical quantities. As illustrated by the studies [23, 24], the streak lines roll up in a false manner in a flow that shows the presence of discrete vortices. Alternatively, the choice of the reference frame has immense importance as steady flows can appear unsteady in an incorrect frame of reference. The identification of the vortices in a vector field along with their diameter and strength calculations is normally accomplished by identifying regions of isolated nature, i.e. the regions containing higher strength of vorticities. In simple flow fields this may be an easy task, but in complex flow regimes this is not an easy task. In such a case, the vortices within a flow regime may be masked by the shearing flow profiles, thus making it difficult to identify the vortices and then to identify their contribution to the overall flow. The method to identify the state of motion of a fluid regime depends on the critical point estimation of the local velocity gradient tensor and its corresponding eigenvalues [25–28]. In a 3D domain, the velocity gradient tensor will have one real eigenvalue (λr) and a pair of complex conjugate eigenvalues (λcr ± iλci). Now, here two conditions can be raised, first when the discriminant of the equation is positive then the particles exhibit a swirling motion. On the other hand, when it is equal to zero, then the path of the particles is infinitely elliptical, and the flow would be the shearing flow. Thus, the vortex study using data of the swirling strength is independent of the frame of reference as it does not give any information related to the regions having the maximum vorticity, but these represent the regions that lack any swirling motion.

The swirling strength and the vorticity has been calculated using the relations $\Omega_{ci} = \lambda_{ci}\varphi_\theta/|\varphi_\theta|$ and $\psi_\theta = \frac{\partial U_x}{\partial r} - \frac{\partial U_r}{\partial x}$ [29]. The method for the calculations related to the swirling strength is the most promising method when dealing with turbulence [30–36]. This method has been developed with the passage of time [37–38]. The swirling strength has been quantified as the imaginary part of the Eigenvalues. These eigenvalues are related to the velocity tensor [39], thus used for the vortex indicator in wall turbulence for strong mean shear indication. The normalized velocity has been acquired by using the axial velocity in denominator, i.e. $\frac{\omega^2}{U_m^2}$.

The occurrence of the vortices across the PIV plane has been estimated by using the MATLAB Blob analysis [40]. As the occurrence of the vortices is mainly clues towards the regions where the vortices inhibited for which the probability density function analysis has been performed. Thus not only their occurrence has been ensured but also the regions have been identified where the maximum probability density function analysis (PDF) has been performed. The details related to the occurrence of the vortices measurements using the Blob analysis has been given as follows.

This technique has been used to identify the regions of the consistent images inside the fluid regions. In this technique the back colors have been classified first, Thus the consistent images have been used to show the strength of the vortices. The combined histogram build at the end has given us the maximum value Cmax with the range of the output mentioned as D [42].

$$C_j = max\left(\frac{w_m M_j - w_B B_j}{C_{max}}, 0\right) \times D \quad (5)$$

The pixels have been given the values based on the RGB color classifications, with the original RGB is being normalized to respective spaces of colors.

$$r = \frac{R}{R+G+B}, g = \frac{G}{R+G+B}, \& b = \frac{B}{R+G+B} \tag{6}$$

The central position (x_c, y_c) The vi which is the velocity for each blob has been calculated as;

$$x_c = \frac{M_{10}}{M_{00}}, y_c = \frac{M_{01}}{M_{00}} \tag{7}$$

$$v_i = \sqrt{(x_c(i) - x_c(i-1))^2 + (y_c(i) - y_c(i-1))^2} / \Delta T \tag{8}$$

whereas;

$$\begin{aligned} M_{00} &= \sum\sum_{xy} I(x, y), \\ M_{10} &= \sum\sum_{xy} xI(x, y), \\ M_{01} &= \sum\sum_{xy} yI(x, y), \end{aligned} \tag{9}$$

With $M_{00}, M_{10}, \&M_{01}$ are the moments with the highest order which were calculated from the image intensity $I(x, y)$. The moment with the order 2 has been calculated as;

$$\begin{aligned} M_{20} &= \sum\sum_{xy} x^2 I(x, y), \\ M_{02} &= \sum\sum_{xy} y^2 I(x, y), \end{aligned} \tag{10}$$

Which helps us to calculate the axis of the blob;

$$\begin{aligned} a &= \sqrt{6\left(p + r + \sqrt{q^2 + (p-r)^2}\right)}, \\ b &= \sqrt{6\left(p + r - \sqrt{q^2 + (p-r)^2}\right)} \end{aligned} \tag{11}$$

whereas;

$$p = \frac{M_{20}}{M_{00}} - x_c^2, q = 2\left(\frac{M_{20}}{M_{00}} - x_c y_c\right), r = \frac{M_{02}}{M_{00}} - y_c^2 \tag{12}$$

Thus, the relative abundance of vortical shapes have been calculated using these relations.

3 Results and Discussion

In the current theoretical and experimental study, steam jet has impinged on a static wall, the resulting flow profiles, the vortices generation and their interaction, along with the area under their influence and their swirling strengths have been obtained with the help of the data acquired by using the PIV System for velocity characterization.

3.1 Analysis of the Vortices

Our understanding regarding the submerged jet injection and the resultant vortices has changed to a large extent. In the previous study [43], the connection of the turbulence with the vortices and their induced entrainment of the surrounding fluid has been discussed. The creation of the turbulent layer comes with the overlapping of a fluid layer over another layer. With these like orientation one layer is the turbulent layer and the other layer is the non-turbulent layer. These layers have relatively at simple configuration with each other and they have their own boundary layers, based on the structures of the interacting fluids inside them. Thus, the vortices that have been created due to the shear of the turbulent high velocity fluid layer can be understand by the propagation for one layer of the fluid inside another layer of the fluid. In such scenarios, the random propagation of the layers into one another in diffusive way into the surrounding fluid layer resulting into the nibbling of the surrounding fluids, The layer with less turbulence placed in surrounding fluid layers have been characterized by the conditions at the boundaries of this layer [44, 45]. This resulted into the creation of the vortices across the interface, as well as surrounding fluids have been entrained. This whole sequence has three sections whereas in the first section, the non-turbulent layer and the middle layer has set into motion by the main layer of the fluid which in its core is turbulent. This resulted into the entrainment and the vortices formation into the surrounding layers. The top second layer comes into contact with the main core fluid layer in the second section, thus the thickness of these two layers then starts decreasing at the smallest Kolmogorov scale. Then in the last stage the mixing will be inhibited and the whole fluid came into motion under the influence of the central core jet.

There are several retarding mechanisms when the jet interacts with the wall. When the supersonic steam jet has impinged on a wall (that include the jet quenching by the wall as well as the retardation being offered to the jet by the surrounding fluids. Normalized vorticity and its swirling strength have been determined and plotted at different inlet pressures (1.5–3.0 bars) as shown in Fig. 3(a–h).

It should be noted that these plots have shown the intermediate stages of the flow in between the events when the steam has been injected and the event when the supply of the steam has been terminated at the start and the end of the data acquisition phase respectively.

It has been observed that at the central region of the plane, the normalized negative vortices have maxima whereas this behavior degrades as soon as the jet moves outward along the radial direction, after the jet striking the wall. This has been shown by the circular marks in Fig. 3(a–h). A possible reason for such a behavior may be that the fluid at high velocity induced vortical structures at its interface with the surrounding water. These vortical structures have positive vorticity with respect to the negatively

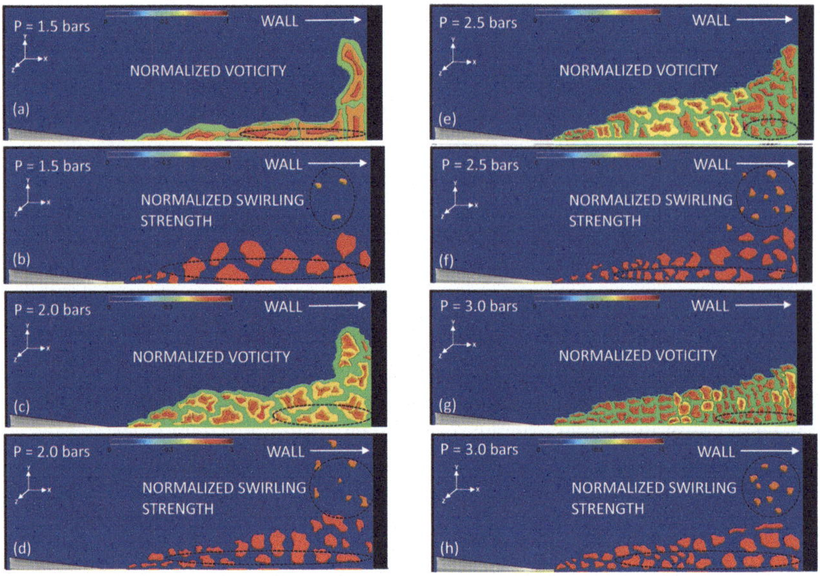

Fig. 3. Normalized vorticity and Normalized Swirling Strength at the varying inlet total pressure 1.5–3.0 bars.

vorticity fluid elements at its interface. These negative vortical structure fluid elements have more concentration at the center which gradually reduce as the vortical structure fluid elements move away from the central portion.

The vortical structures and their swirling strengths have been determined using the relations given before. The vortical structures are present both in the jet region as well as near the wall region. However, at lower inlet pressure, a clear indication occurs towards the existence of the Kelvin Helmholtz instabilities, which act within the fluid layers resulting in the vorticities that rotate in anti-clockwise directions. These are the vortical structures encapsulating the round jet, which results in sustained enveloping profiles around the jet which afterwards break near the wall into small vortical structures and being expelled radially outwards. The distance, by which the enveloping vortical structures encapsulating the main round body of the jet, ranges from 0 D to 3.0 D. This is the standoff distance at which the nozzle is placed away from the wall.

The enveloping vortical structures encapsulate the round body of the jet before they strike with the wall, resulting in the structures quickly spreading out due to the radially expanding shear layer near the wall, this then tears them into small vortical structures moving away from the striking/impingement point. These are the secondary vortices (shown by the orange color contours in each figure). Large secondary vortices don't appear across the region when the jet reaches to impinge at the wall. These large vortices are not clearly detected in the region even near the wall, rather a densely populated region (secondary vortices) near the wall will give rise to the vortices of different sizes and shapes across this chaotic region.

3.2 Transformation from Negative to Positive Vortices and Their Spatial Occurrence Probability

The formation of the vortical structures inside the two-phase flows and their transformation from negative to the positive vortices or vice-versa has a lot of implications in the fluid dynamic studies of multiphase flows. Starting from the propagation of the shock-waves in two-phase flows [46] to fast growing droplets in clouds, where the clustering of such vortices resulted into an increased concentration of the water contents after collisions among cloud's entities [11, 12] etc. The vorticities and their swirling strengths indicate the most traveled tracks and the shearing layers as shown in Fig. 4(a–c).

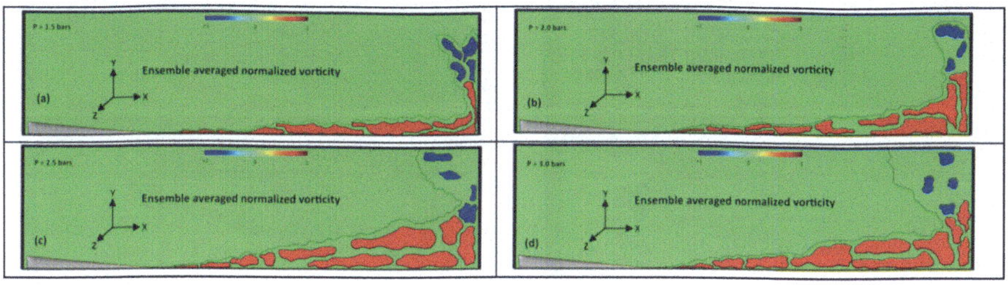

Fig. 4. (a-d) Ensemble average vorticity profiles at the varying inlet pressure (1.5–3.0 bars)

The clockwise vortical structures have been transformed into a thick layer of fluid vorticities which will be generated due to the turbulence induced by the interaction of the primary structures with the wall. The conversion from the primary to the secondary vortices mostly happened along the direction perpendicular to the axial axis, at the distance ranging from r = 10–30 mm (Fig. 5(a–d)). It should be noted that in Fig. 5(a-d) the normalized vorticity has been shown at the start of the process. Whereas, just few seconds before the end of the process the normalized vorticity contours have been shown in Fig. 5(e–h).

The normalized vorticity contours and the density of the transformed positive vortices is very low at the start of the process. At the end of the process, the density of these positive vortices is considerable as shown in Fig. 5(e–h).

These vortices exist around the nozzle exit and, in the region, where enveloping vortices are generated all the way to the region near the wall. These positions along with the axial axis range from X/D = 0 till ~ 3 corresponding to the inlet pressure, P = 1.5–3.0 bars.

Along the normalized radial distance (r/D), the profiles for the ensembled averaged swirling strengths were estimated from r/D = 0 to r/D = 15. In this regard, to clarify the existence of the primary and secondary vortices, and their spatial transformation from primary to secondary vortices, Probability Density Function (PDF) analysis has been performed to calculate and find the regions where the maximum vorticities exit, mainly depending on the clock and anti-clockwise rotations. We have determined the points around which the maximum density of such vortices exits and their strength [41] and their abundance has been calculated and plotted as shown in Fig. 7(a and b).

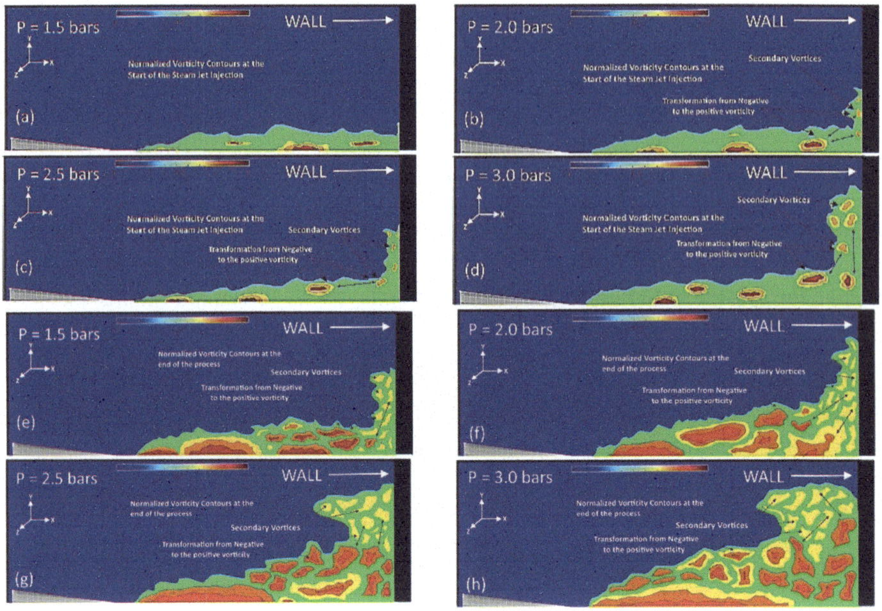

Fig. 5. (a-d) Transformation of negative to positive vorticity after impingement of ths steam jet with the wall (at the start of the process). (e-h) Transformation of negative to positive vorticity after impingement of the steam jet with the wall (at the start of the process). The profiles for the normalized swirling strengths can be seen in Fig. 6(a–d), which indicate the strongest vortices reside within the fluid domain.

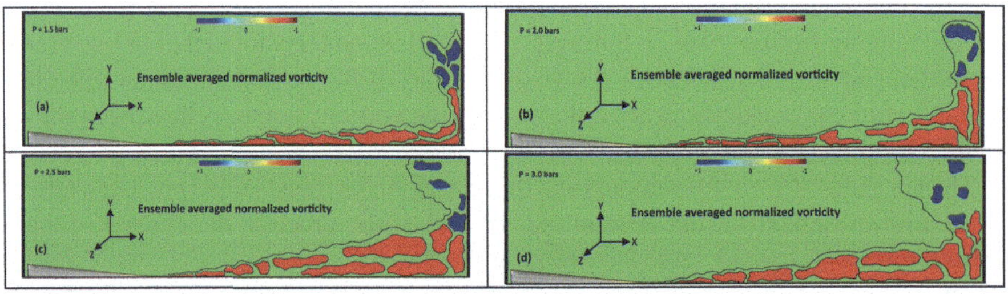

Fig. 6. (a-d) Ensemble average swirling strength at varying pressure ranges from 1.5–3.0 bars.

From Fig. 7(a & b), With the rise in the inlet steam pressure, there is a slight increase in the number voritices that occurred initially, in the surroundings of the main jet region. However, in the region near the wall, a small increase in the density of the dimensionless secondary vortices is observed. These results have been supported by an increasing trend in the area which has the impact of the primary and secondary vortical structures. The normalized vortical structures density for the primary vortices have shown an increasing trend in terms of their number density that ranges from 0.2 to 0.6 for primary vortices, and from 0 to 0.3 for secondary vortices. Whereas in terms of the normalized area ratio,, the range stretches between 0.02 and 0.6 for primary vortices and 0.01 and 0.3 for secondary

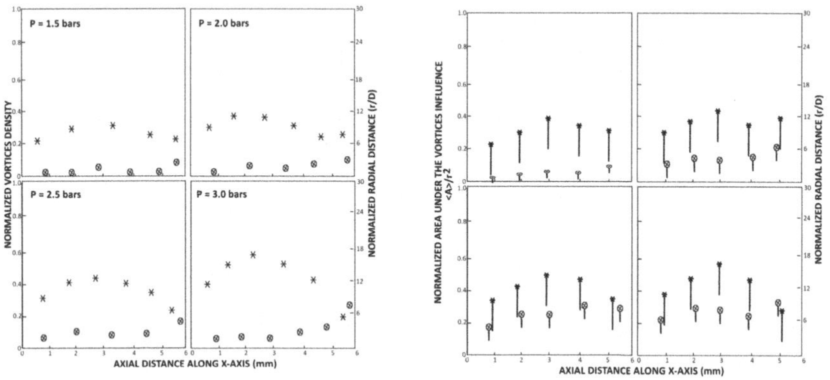

(a) (b)

Fig. 7. **(a).** Normalized Vortices density (⋆ Primary vortices) & (⊗ Secondary vortices) across axial and dimensionless radial directions at varying pressure (1.5–3.0) bars. **(b).** Normalized Area under the vortical influence (⍏ Primary vortices) & (⊗ Secondary vortices) across axial and dimensionless radial directions at varying pressure (1.5–3.0) bar

vortices within the surroundings of the jet and in the near-wall region respectively. The interaction region and region near the wall following the impingement are characterized by the length (X = 5-6mm), which corresponds to the stand-off distance between the supersonic steam jet nozzle and the distance that has been normalized (r/D) that ranges from 0–30. Whereas the normalized primary vortical structure density peaks exist at 0.27 and the secondary vortical structures density peaks exist at 0.22. Also, the normalized region which has been dominantly under the influence of the vortical structures peaks exist at 0.4 and the area under the influence of the secondary vortices structures peaks exist at 0.3 in these regions. Furthermore, in these regions, in most of the cases, the normalized vortical structure density and the region under the impact of the structures are found to be decreasing. However, this slight decrease may be attributed to the vorticity diffusion [47].

4 Conclusion

In the current manuscript, an experimental study has been undertaken, involving impingement of a steam jet onto a wall. The characterization of jet's impingement is based on the generation of vortices as well as their normalized swirling strengths. It has been found that with pressure from 1.5–3.0 bars, the normalized vortical structures density distribution has spread widely, which gives birth to the region of extremely chaotic activities. Primary vortices have been transformed into secondary vortices within the impingement region and in the region near the wall, where they are found to be expelled outward along the radial direction after the impingement. The density distribution of the normalized vortical structure through the spatial Probability Density Function (PDF) analysis reveals that the maximum number of secondary vortices are born near the interaction region with the wall, along the radial directions, whereas, the probability of such vortices being detected within the jet region is small, which may be attributed to the diffusion of vortices into each other.

Acknowledgment. The authors are thankful to the Ministry of Science and Higher Education of the Russian Federation for supporting this work (through project No. FENU-2023–0010) for their support to this work.

References

1. Bakke, P.: An experimental investigation of a wall jet. J. Fluid Mech. **2**, 467–472 (1957). https://doi.org/10.1017/S0022112057000270
2. Deng, T., Norris, S., Sharma, R.N.: Numerical investigation on the stability of tunnel smoke stratification under the effect of water spray and longitudinal ventilation. Tunn. Undergr. Sp. Technol. **112**, 103901 (2021). https://doi.org/10.1016/J.TUST.2021.103901
3. Lede, J., Villermaux, J., Ouzane, R., Hossain, M.A., Ouahes, R.: Production of hydrogen by simple impingement of a turbulent jet of steam upon a high temperature zirconia surface. Int. J. Hydrogen Energy **12**, 3–11 (1987). https://doi.org/10.1016/03603199(87)90120-0
4. Li, X., Gaddis, J.L., Wang, T.: Modeling of heat transfer in a mist/steam impinging jet. J. Heat Transf. **123**, 1086–1092 (2001). https://doi.org/10.1115/1.1409262
5. Domnick, C.B., Bera, F.K., Brillert, D., Musch, C.: Modification of a steam valve diffuser for enhanced full load and part load operation using numerical methods, Period. Polytech. Mech. Eng. **60**, 185–192 (2016). https://doi.org/10.3311/PPme.9041
6. Woisetschläger, J., Jericha, H., Sanz, W., Gollner, F.: Optical investigation of transonic wall-jet film cooling. ASME Cogen Turbo Power **95** (1995)
7. Shademan, M., Balachandar, R., Roussinova, V., Barron, R.: Round impinging jets with relatively large stand-off distance, Phys. Fluids **28** (2016). https://doi.org/10.1063/1.4955167
8. Launder, B.E., Rodi, W.: The turbulent wall jet. Prog. Aerosp. Sci. **19**, 81–128 (1979). https://doi.org/10.1016/0376-0421(79)90002-2
9. Chung, Y.M., Luo, K.H.: Unsteady heat transfer analysis of an impinging jet. J. Heat Transf. **124**, 1039–1048 (2002). https://doi.org/10.1115/1.1469522
10. Hoogendoorn, C.J.: The effect of turbulence on heat transfer at a stagnation point. Int. J. Heat Mass Transf. **20**, 1333–1338 (1977). https://doi.org/10.1016/0017-9310(77)90029-1
11. Lee, J., Lee, S.J.: The effect of nozzle configuration on stagnation region heat transfer enhancement of axisymmetric jet impingement. Int. J. Heat Mass Transf. **43**, 3497–3509 (2000). https://doi.org/10.1016/S0017-9310(99)00349-X
12. Khan, A., Sanaullah, K., Sobri Takriff, M., Hussain, A., Shah, A., Rafiq Chughtai, I.: Void fraction of supersonic steam jet in subcooled water. Flow Meas. Instrum. **47**, 35–44 (2016). https://doi.org/10.1016/J.FLOWMEASINST.2015.12.002
13. Khan, A., et al.: Numerical and experimental investigations on the physical characteristics of supersonic steam jet induced hydrodynamic instabilities. Asia Pac. J. Chem. Eng. **11**, 271–283 (2016). https://doi.org/10.1002/APJ.1963
14. Khan, A.: CFD based hydrodynamic parametric study of inclined injected supersonic steam into subcooled water (2014). https://doi.org/10.3850/978-981-09-4587-9_P03
15. Khan, A., Sanaullah, K., Takriff, M.S., Zen, H., Fong, L.S.: Inclined injection of supersonic steam into subcooled water: a CFD analysis. Adv. Mater. Res. **845**, 101–107 (2014). https://doi.org/10.4028/WWW.SCIENTIFIC.NET/AMR.845.101
16. Khan, A., Sanaullah, K., Haq, N.U.: Development of a sensor to detect condensation of supersonic steam. Adv. Mater. Res. **650**, 482–487 (2013). https://doi.org/10.4028/www.scientific.net/AMR.650.482

17. Yule, A.J.: Large-scale structure in the mixing layer of a round jet. J. Fluid Mech. **89**, 413–432 (1978). https://doi.org/10.1017/S0022112078002670
18. Nishino, K., Samada, M., Kasuya, K., Torii, K.: Turbulence statistics in the stagnation region of an axisymmetric impinging jet flow. Int. J. Heat Fluid Flow **17**, 193–201 (1996). https://doi.org/10.1016/0142-727X(96)00040-9
19. Zuckerman, N., Lior, N.: Radial slot jet impingement flow and heat transfer on a cylindrical target. J. Thermophys. Heat Transf. **21**, 548–561 (2007)
20. Khan, A., et al.: Pressure stresses generated due to supersonic steam jet induced hydrodynamic instabilities. Chem. Eng. Sci. **146**, 44–63 (2016). https://doi.org/10.1016/j.ces.2016.01.056
21. Shah, A., Chughtai, I.R., Inayat, M.H.: Numerical simulation of direct-contact condensation from a supersonic steam jet in subcooled water. Chin. J. Chem. Eng. **18**, 577–587 (2010). https://doi.org/10.1016/S1004-9541(10)60261-3
22. Entry, E.B.: Steam tables, A-to-Z guid. to Thermodyn. Heat Mass Transf. Fluids Eng. (2011). https://doi.org/10.1615/ATOZ.S.STEAM_TABLES
23. Hama, F.R.: Streaklines in a perturbed shear flow. Phys. Fluids **5**, 644–650 (1962). https://doi.org/10.1063/1.1706678
24. Gursul, I., Lusseyran, D., Rockwell, D.: On interpretation of flow visualization of unsteady shear flows. Exp. Fluids **9**, 257–266 (1990). https://doi.org/10.1007/BF00233126
25. Chong, M.S., Perry, A.E., Cantwell, B.J.: A general classification of three-dimensional flow fields. Phys. Fluids A **2**, 765–777 (1990). https://doi.org/10.1063/1.857730
26. Dallmann, U.: On the footprints of three-dimensional separated vortex flows around blunt bodies. AGARD CP **494** (1991)
27. Zhou, J., Adrian, R.J., Balachandar, S.: Autogeneration of near-wall vortical structures in channel flow. Phys. Fluids **8**, 288–290 (1996). https://doi.org/10.1063/1.868838
28. Zhou, J., Adrian, R.J., Balachandar, S., Kendall, T.M.: Mechanisms for generating coherent packets of hairpin vortices in channel flow. J. Fluid Mech. **387**, 353–396 (1999). https://doi.org/10.1017/S002211209900467X
29. Adrian, R.J., Christensen, K.T., Liu, Z.C.: Analysis and interpretation of instantaneous turbulent velocity fields. Exp. Fluids **29**, 275–290 (2000). https://doi.org/10.1007/s003480000087
30. Herpin, S., Stanislas, M., Soria, J.: The organization of near-wall turbulence: a comparison between boundary layer SPIV data and channel flow DNS data. J. Turbul. **11**, 1–30 (2010). https://doi.org/10.1080/14685248.2010.508460
31. Lee, S.J., Choi, Y.S.: Decrement of spanwise vortices by a drag-reducing riblet surface. J. Turbul. **9**, 1–15 (2008). https://doi.org/10.1080/14685240802251517
32. Natrajan, V.K., Wu, Y., Christensen, K.T.: Spatial signatures of retrograde spanwise vortices in wall turbulence. J. Fluid Mech. **574**, 155–167 (2007). https://doi.org/10.1017/S0022112006003788
33. Wu, Y., Christensen, K.T.: Population trends of spanwise vortices in wall turbulence. J. Fluid Mech. **568**, 55–76 (2006). https://doi.org/10.1017/S002211200600259X
34. Hambleton, W.T., Hutchins, N., Marusic, I.: Simultaneous orthogonal-plane particle image velocimetry measurements in a turbulent boundary layer. J. Fluid Mech. **560**, 53–64 (2006). https://doi.org/10.1017/S0022112006000292
35. Christensen, K.T., Wu, Y.: Visualization and characterization of small-scale spanwise vortices in turbulent channel flow. J. Vis. **8**, 177–185 (2005). https://doi.org/10.1007/BF03181661
36. Christensen, K.T., Adrian, R.J.: Statistical evidence of hairpin vortex packets in wall turbulence. J. FluidMech. **431**, 433–443 (2001). https://doi.org/10.1017/S0022112001003512
37. Tomkins, C.D., Adrian, R.J.: Spanwise structure and scale growth in turbulent boundary layers. J. Fluid Mech. **490**, 37–74 (2003). https://doi.org/10.1017/S0022112003005251

38. Chen, Q., Adrian, R.J., Zhong, Q., Li, D., Wang, X.: Experimental study on the role of spanwise vorticity and vortex filaments in the outer region of open-channel flow. J. Hydraul. Res. **52**, 476–489 (2014). https://doi.org/10.1080/00221686.2014.919965
39. Chen, H., Li, D., Bai, R., Wang, X.: Comparison of swirling strengths derived from two- and three-dimensional velocity fields in channel flow. AIP Adv. **8**, 055302 (2018). https://doi.org/10.1063/1.5023533
40. MATLAB. Statistics for labeled regions - Simulink (n.d.)
41. Panton, R.L.: Incompressible flow (n.d.)
42. Lu, S., Tsechpenakis, G., Metaxas, D.N., Jensen, M.L., Kruse, J.: Blob analysis of the head and hands: a method for deception detection. In: Proceedings Annual Hawaii International Conference on System Science, p. 20 (2005). https://doi.org/10.1109/hicss.2005.122
43. Corrsin, S., Kistler, A.L.: Free-Stream Boundaries of Turbulent Flows. Johns Hopkins University, Balitmore (1955)
44. Tennekes, J.L.H.: A First Course in Turbulence - Emeritus Professor of Aeronautical Engineering. MIT Press (1972)
45. Clauser, F.H.: The structure of turbulent shear flow. Nature **179**, 60 (1957). https://doi.org/10.1038/179060a0
46. Smirnov, N.N., Zverev, N.I., Tyurnikov, M.V.: Two-phase flow behind a shock wave with phase transitions and chemical reactions. Exp. Therm. Fluid Sci. **13**, 11–20 (1996). https://doi.org/10.1016/0894-1777(95)00171-9
47. van Hout, R., Rinsky, V., Grobman, Y.G.: Experimental study of a round jet impinging on a flat surface: Flow field and vortex characteristics in the wall jet. Int. J. Heat Fluid Flow **70**, 41–58 (2018). https://doi.org/10.1016/j.ijheatfluidflow.2018.01.010

Intelligent Unmanned System

Real and Predicted Temperature Distributions of LEO Satellite Attitude Determination and Control System (ADCS)

Nayef Alshamlan, Majed A. Alharbi(✉), Anas Alburayt, Yasser H. Alattas, and Mohammed A. Alturky

King Abdulaziz City for Science and Technology (KACST), Space Technologies Institute (STI), Riyadh 11442, Saudi Arabia
{nalshamlan,mbharbi,abareet,yalattas,malturky}@kacst.gov.sa

Abstract. The Attitude Determination and Control System (ADCS) plays a crucial role in ensuring the stability and accuracy of satellite orientation in the presence of external disturbance torques. ADCS is one of the satellite subsystems comprising a combination of actuators and sensors that work together to regulate and align the satellite according to the desired concept of operations. ADCS system relies on external references to determine the satellite's angular orientation that is relative to a fixed inertial reference frame. It is important to note that each ADCS system is designed to fulfill the specific requirements of its parent satellite, resulting in unique configurations and capabilities. The design and implementation of an ADCS system necessitates careful consideration of the satellite's operational needs and the external disturbances it will encounter. By effectively managing these factors, the ADCS system ensures the precision and stability of the satellite's orientation throughout its mission lifespan. In the space environment, the satellite subsystems are exposed to solar radiation (heat flux) and heat generated from the satellite components. This paper presents a thermal analysis case study of ADCS from an ongoing KACST low-earth-orbit (LEO) mission to guarantee that the system is kept within the operating temperature range. Numerical simulation using finite element modeling and analysis (FEMAP) is used. Different thermal scenarios are simulated at nominal and extreme (cold and hot) conditions are considered. Results show that the minimum and maximum average temperatures experienced by the ADCS during orbital operation are 7.4 and 30.8 °C, with a nominal range of 15 and 21.3 °C, respectively. Finally, the nominal predicted results are compared with a measured in-orbit result, with ranging temperatures of 14.3 and 16.3 °C. The ADCS nominal real-time measurements from the ongoing mission fall within the predicted, and as a result, the design accepted temperature range.

Keywords: LEO · ADCS · Thermal Analysis

© The Author(s), under exclusive license to Springer Nature Singapore Pte Ltd. 2025
J. Mo and P. Chutima (Eds.): ICMSC 2024, LNME, pp. 125–131, 2025.
https://doi.org/10.1007/978-981-96-5380-5_12

1 Introduction

Satellite operations in space are generally complex and are mainly sensitive to orbital stability and positioning. The need for communication and modern imaging/surveying applications require directional accuracy and specific object targeting, which are carried out by the ADCS [1]. In other words, the system's main objective is to control and determine the attitude and orientation of the spacecraft utilizing surrounding references such as the Sun, Earth's horizon, the local magnetic field, and the stars, with respect to a determined reference frame [2]. As a satellite fundamental part that deals with delicate control operations within the harsh space environment, it is important to study the potential effects of motion and temperature variations on the ADCS during a LEO mission [3]. In this paper, we will strictly focus on the thermal verification of the subsystem. In space, the spacecraft and its systems are sensitive to external (solar) and internal (satellite components) thermal radiative exposure [4]. Grasping how the ADCS in a satellite responds to heat is vital for enhancing its function and maintaining its safety. Models, simulations, and applications of small satellite thermal analysis were reported by Shen et al. [5]. They reviewed the development of satellite thermal model techniques, parameters, thermal modeling methods and applications. This research investigates how the conditions of outer space affect the temperature of satellite ADCS and examines the temperature differences across a nominal solar day. It then highlights and contrasts the actual temperature readings with those predicted by a computer model. The paper consists of the following sections: An analysis part covering the system structure and thermal model, the numerical simulation, the orbit definition, and the thermal test. This is followed by the results and discussion section, and is then concluded with some observed remarks.

2 Analysis

2.1 Structure and Thermal Model

The configuration of the satellite is outlined in a three-dimensional representation to adhere to the specifications of the production procedure and inspection. Utilizing a three-dimensional model allows for the retrieval of data regarding the components of the satellite for the intent of simulation and assessment. The illustration in Fig. 1 exhibits the three-dimensional thermal depiction of the satellite through the utilization of FEMAP software version 11.3 as the computational tool for all evaluations [6].

2.2 Numerical Simulation

In the current investigation, mainly the ADCS of the satellite will be the focal point. Utilizing the FEMAP software, the study conducts simulations to contrast the temperatures of ADCS under normal and extreme thermal circumstances, which represent potential scenarios of utmost severity. The standard scenario employs mean values for various parameters such as solar radiation. These simulations encompass a full solar day (24 h) to encompass all significant power fluctuations encountered by the ADCS [6]. The analysis cases are outlined in Table 1.

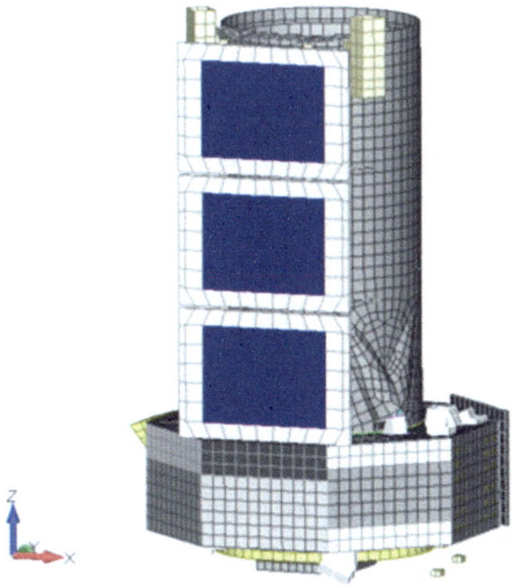

Fig. 1. The satellite 3D thermal model.

Table 1. Analysis Cases.

Parameter\Case	Nominal	Hot	Cold
Time of year	Average solar flux	December Solstice	June Solstice
Solar flux	1368.6 W/m^2	1411.6 W/m^2	1323.6 W/m^2
Albedo	0.306	0.35	0.24
Earth IR	237 W/m^2	263 W/m^2	224 W/m^2
Internal dissipation	Full Performance	Full Performance	Housekeeping
Thermo-optical properties	Beginning of life	End of life	Beginning of life

2.3 Orbit Definition

A mission-representative orbit, as described in Table 2, is used for the analysis.

Table 2. Orbit Definition.

Parameter	Value
Type	Sun-synchronous
Altitude	547 km
Eccentricity	0°
Orbit inclination	97.63°
Orientation	Primary: −Y aimed at Sun Secondary: +Z aligned with Nadir

2.4 Thermal Test

A thermal design test is performed to maintain the satellite components' temperatures within its range. It is essential for both the physical integrity of the satellite and for its efficient operation. The electronic equipment needs to have their optimum performance within a certain temperature range. The satellite ADCS box temperatures were measured to ensure that they are compatible with the space environment and their values are within an accepted tolerance range of -1 to 44 °C and a required design range of 14 to 28 °C, respectively [7].

3 Results and Discussion

Thermal analysis was conducted on all satellite subsystems. The current study focuses solely on the simulated temperature distribution of the satellite ADCS box over a 24-h solar day, illustrated in Figs. 2, 3 and 4 for the nominal, cold, and hot scenarios. It is important to mention that the temperatures predicted are illustrated by minimum and maximum curves, with the minimum curves representing when the satellite is in the shadow (not facing the sun) and the maximum curves indicating sun-facing conditions. The case study aims is to highlight the nominal temperature values of the satellite ADCS by comparing the in-orbit temperatures obtained from the ground station with the predicted temperatures. Extreme worst-case thermal conditions are simulated to establish upper and lower bounds for reference. For the extremely hot scenario, the temperature ranges are 25 and 30.8 °C, shown in Fig. 2 due to high heat fluxes and system operation. On the other hand, the cold scenario reaches temperature ranges of around 7.4 and 10.2 °C, as illustrated in Fig. 3, due to low heat fluxes and minimal system operation. The focus of this paper is on the results of the nominal thermal case, which represents

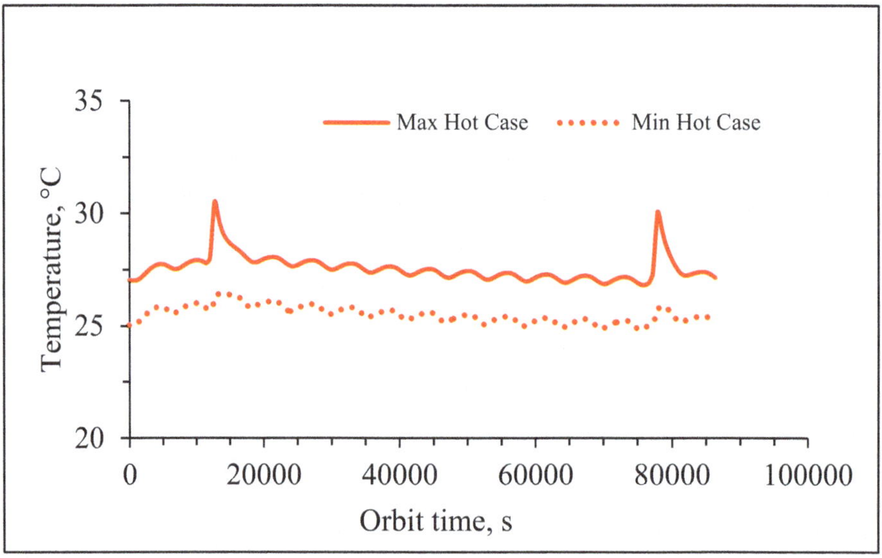

Fig. 2. Predicted temperatures of the ADCS (Hot).

the primary operating scenario of the satellite throughout its lifespan; The temperature ranges experienced by the ADCS box during nominal orbital operation are 15 and 21.3 °C, as shown in Fig. 4. Whereas, Fig. 5 displays the ADCS temperature distribution ranges obtained from the ground station, with temperature values ranging between 14.3 to 16.3 °C. Consequently, all actual and predicted temperature values fall within the design range (14–28 °C), when compared to the operational temperatures measured [7].

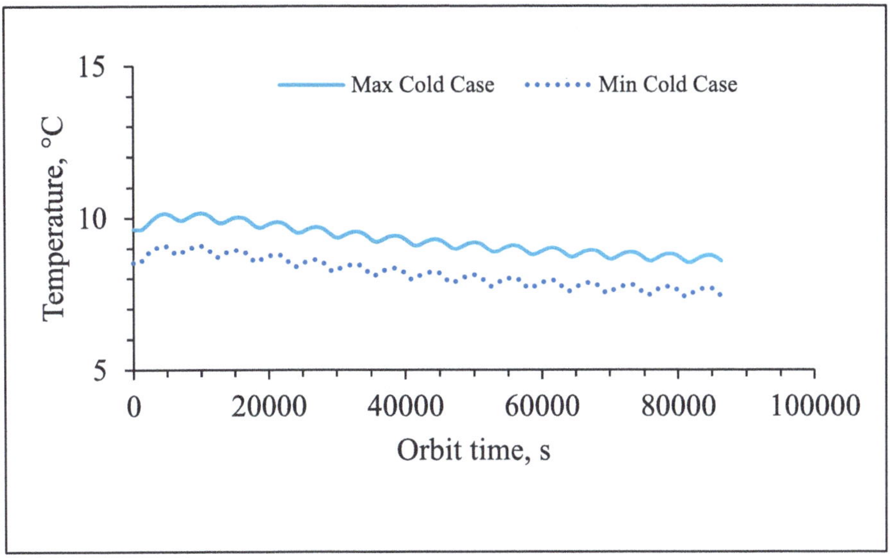

Fig. 3. Predicted temperatures of the ADCS (Cold).

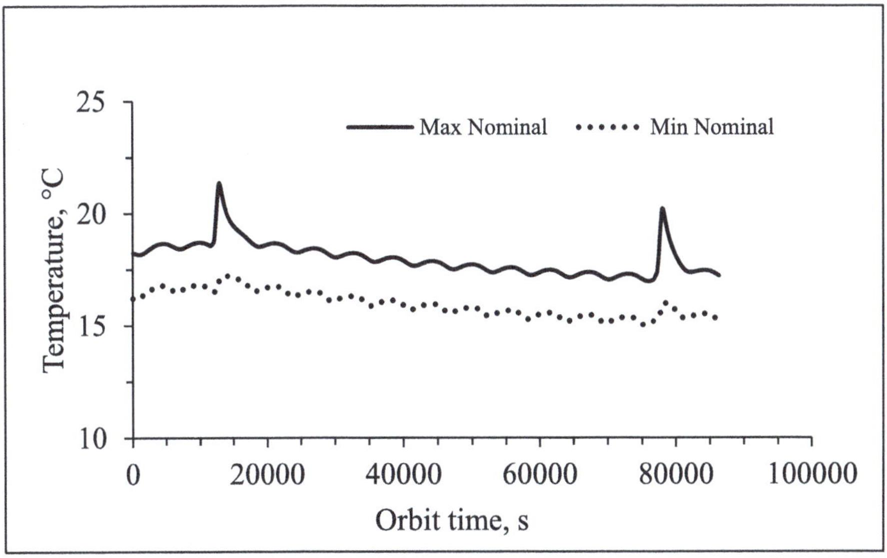

Fig. 4. Predicted temperatures of the ADCS (Nominal).

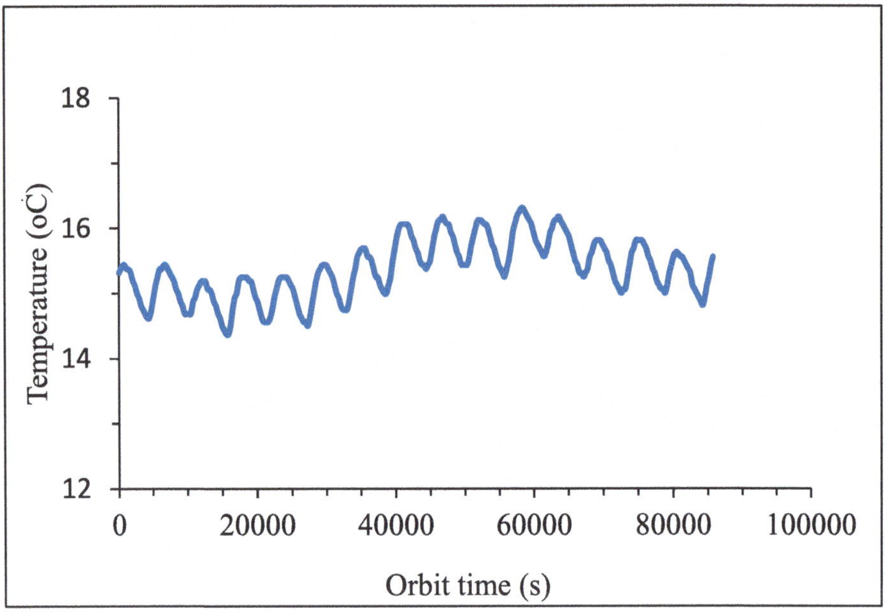

Fig. 5. Mission recorded temperatures of the ADCS (Nominal).

4 Conclusion

To evaluate the temperature performance of satellite ADCS in orbit, thermal design and simulation were carried out utilizing FEMAP software. Present were both the predicted and actual in-orbit temperatures. Nominal, hot, and cold (for reference) scenarios were simulated. The predicted values and the actual captured temperature values fell within the accepted design requirement with a stable average in-orbit temperature of 15.3 ± 1 °C during an average solar day. The average temperature value can be attributed to suitable design decisions for the satellite ADCS, as the thermal distribution has a maximum 2 °C variation along its nominal cycle. Any discrepancies, however, between the real and test results can be attributed to several factors, such as minor differences in orbital parameters at launch, in addition to expected contrast between simulated and accessible real in-orbit periods, and any potential external cosmetic phenomena.

Acknowledgments. The authors would like to thank KACST for supporting the research work. This work could not have been carried without the efforts of the STI colleagues, Abdulaziz Alahmadi, Abdulaziz Alharbi, and Ebraheem Alanzi, for providing the ADCS data.

References

1. Bolandi, H., Haghparast, M., Saberi, F., Vaghei, B., Smailzadeh, S.: Satellite attitude determination and control. Meas. Control (45), 151–157 (2012)

2. Steyn, W.: Stability, pointing, and orientation. In: Pelton, J. (ed.) Handbook of Small Satellites. Springer, Cham (2020)
3. Kaplan, M.: Space missions and applications. Launch Space (2011)
4. Gilmore, D.: Spacecraft Thermal Control Handbook, Volume I: Fundamental Technologies (2002)
5. Shen, M., Zhang, L., Sun, Z., Kong, L., Liu, Y., Xue, Z.: Satellite attitude determination and control. Adv. Space Res. **74**(2), 836–860 (2024)
6. Siemens Digital Industries Software. http://blogs.sw.siemens.com/simcenter/femap-version-11-3-2/
7. KACST Ground Station, Riyadh, Saudi Arabia

Development of a Guarded Drone for Safe Flight in Confined Spaces

Mitsuki Uesugi, Jae Hoon Lee(✉), and Shingo Okamoto

Graduate School of Science and Engineering, Ehime University, 3 Bunkyocho, Matsuyama, Ehime 790-0826, Japan
lee.jaehoon.mc@ehime-u.ac.jp

Abstract. Drones have garnered significant attention across various industries, including agriculture, logistics and inspection, due to their ability to operate in confined or elevated spaces that are inaccessible to human workers. However, drones flying in indoor spaces present unique challenges, particularly the risk of collisions with walls and obstacles that can lead to crashes. To address this issue, this study focuses on the development of a protective guard designed to enclose the entire drone, thereby preventing contact with obstacles and reducing crash risks. The guard is constructed using lightweight and durable Carbon Fiber Reinforced Plastics (CFRP) material, specifically employing thin and flexible CFRP flat bars. The spherical shape of the guard ensures that the drone can maintain flight regardless of the direction of impact, providing omnidirectional protection. Additionally, the design incorporates a low center of gravity, which allows the drone to automatically return to its original orientation for taking off again if it falls. This mechanism enhances the stability and resilience of the drone in indoor environments. The effectiveness of the guard was evaluated through a series of tests, demonstrating its ability to protect the drone during flight, even with light contact with objects.

Keywords: drone · guard · CFRP

1 Introduction

1.1 Background

In recent years, drones have gained attention across various industries such as agriculture, logistics, and inspection [1, 2]. Their use has recently become more common in indoor applications. Since drones can be remotely controlled or fly autonomously, they are expected to operate in elevated or confined areas where humans cannot reach [3]. However, unlike typical drones flying in outdoor free spaces, a drone flying in indoor environments poses a higher risk of contact with walls or obstacles. Therefore, sensors and cameras have been embedded on the drone to avoid collisions with walls and obstacles [4]. Nevertheless, even with sensors and cameras, there is still a risk of collision if objects are not detected correctly or if the drone is flying in extremely narrow spaces. For such situations, if the propellers and parts of the drone are protected with a guard,

© The Author(s), under exclusive license to Springer Nature Singapore Pte Ltd. 2025
J. Mo and P. Chutima (Eds.): ICMSC 2024, LNME, pp. 132–140, 2025.
https://doi.org/10.1007/978-981-96-5380-5_13

it can continue flying even in the case of contact with objects. In order to realize the concept this research aims to develop a lightweight guard that protects the drone from crashing, even if it cannot avoid a collision.

1.2 Research Objectives

This research has two main objectives in developing a drone with guards. These are the continuous flight function and the posture-recovery function. Firstly, the continuous flight function involves designing the guard in a way that covers the entire drone. While most existing studies have guards that cover only particular parts of the drone, i.e. Propellers, the aim of this research is to prevent crashes regardless of the direction of contact with walls or obstacles by covering the entire body of the drone [5]. Secondly, the posture recovery function involves designing a guard structure that allows the drone to return to a flyable state, specifically balanced horizontal pose, after falling. By lowering the center of gravity and designing the guard frame with a spherical shape, the drone rolls naturally to the stable pose where the center of gravity is lowest when the drone is on flat ground.

2 Drone System Configuration

This section describes the basic configuration of the drone developed in this research. The drone is a quadcopter with four propellers as shown in Fig. 1. It consists of a frame, flight controller, ESC (Electronic Speed Controller), motors, propellers, and receiver. The flight controller used in this study includes a power board function. The drone is powered by a 4-cell lithium-ion polymer (LiPo) battery and is equipped with 2300 kV brushless motors paired with 5-inch propellers.

The configuration of the drone's control system is illustrated in Fig. 2, Where the black solid line represents the power supply wire, and blue dashed line represents the signal wire. As the flight controller incorporates the power board function within itself, the battery power is directly connected to the flight controller, which then supplies power to the four ESCs for rotating motors. The drone is operated using manual control method. The motion command of the operator is transferred from the transmitter to the receiver through wireless communication. Then the flight controller computes the rotational speed of all motors based on both the motion command and the drone posture. Consequently, the rotational speed of each motor is sent to the four ESCs which control the motor.

3 Production of Guards Using Styrofoam

3.1 Continuous Flight Function

In this research, the first focus is on the continuous flight function even in some contact with objects, and a prototype of the drone guard was developed in accordance with this objective. In the prototype phase, the drone guard was constructed of polystyrene plate, a cost-effective and readily processable material. The guard's design and shape were improved with consecutive experiments and studies, as shown in Fig. 3 [6]. Where the guard frame was constructed from 3mm-thick polystyrene plate.

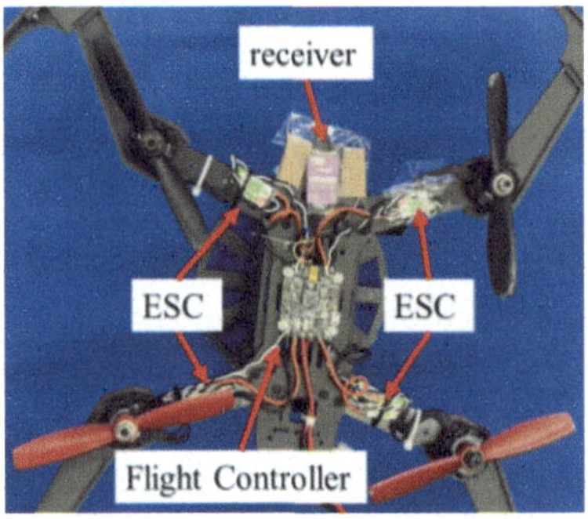

Fig. 1. Drone developed in this research

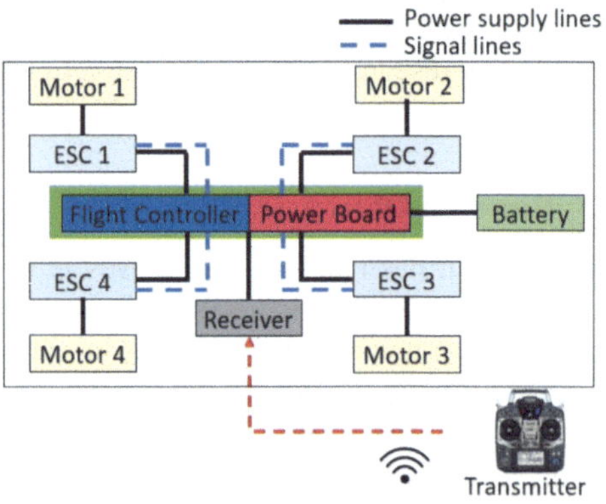

Fig. 2. Configuration of drone control system

To determine the resilience of this prototype, experiments were conducted in which the drone, equipped with the guard frame, was flown and made to contact a wall. Upon contact with the wall, it was observed that the guard frame was not robust, some parts of the guard were bent, and the propeller collided with the guard frame or other objects, causing the drone to crash. To address this issue, additional support columns were incorporated into the design, as displayed in Fig. 4. Another flight and collision experiments were conducted to test the drone equipped with the reinforced guard. As a result, the modified guard showed enhanced resistance to deformation, preventing the propeller from hitting the wall, thereby avoiding a crash.

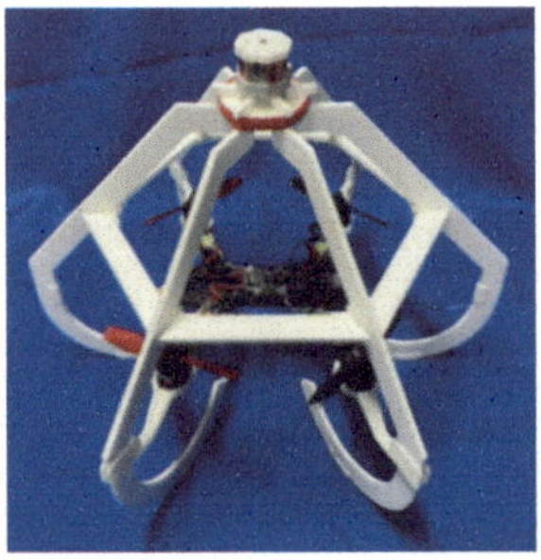

Fig. 3. Drone produced in previous research

Fig. 4. Drone with modified guard frame

3.2 The Posture-Recovery Function

The posture-recovery function was added to allow the guarded drone to resume flight after falling. This function was inspired by the traditional Japanese doll *Okiagari Koboshi,* a traditional self-righting doll, as shown in Fig. 5, and has been adopted into the drone guard design [7]. The center of gravity of the *Okiagari Koboshi* is lowered by placing weights in the bottom. When the doll is tilted, the center of gravity rises, creating a restoring force that returns it to an upright position, which is the equilibrium state where the potential energy of the center of gravity is minimized. To apply this function to a guarded drone, the drone body and battery serve as the main source of weight. In order to lower the center of gravity, the upper part of the guard frame must be designed as vacant space, as displayed in Fig. 6. Additionally, the entire guard was made curved and spherical to allow it to roll easily, regardless of the direction of tilt. This design enables the guarded drone to automatically return to its original posture after rolling on the ground.

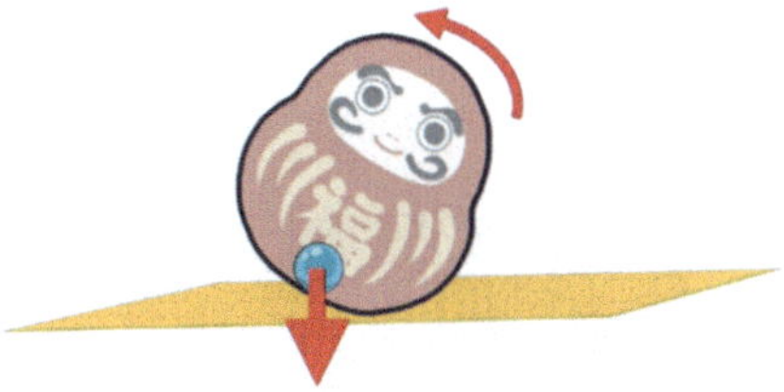

Fig. 5. *Okiagari Koboshi* (self-righting doll) [7]

Fig. 6. Drone with spherical guard frame

4 Production of Guards Using CFRP

4.1 Change of Material

Up to this point, the guard was constructed using polystyrene plate, a material that is easy to process, and the continuous flight function and the posture-recovery function were evaluated. However, the polystyrene guard is weak and susceptible to damage from repeated impacts with walls and landings, ultimately leading to destruction. Therefore, it is necessary to change the material of the guard to protect the drone more securely. When choosing material, it is important to consider the weight of the drone and its negative effect on battery consumption, which in turn leads to a reduction in flight time. To address this issue, it was decided to use CFRP, a material that is both lightweight and durable.

4.2 Change of Shape

With the change of material from polystyrene plate to CFRP, the structure of the guard was also redesigned. The new structure was inspired by a pendant light, as shown in Fig. 7, where thin bamboo rods cover the bulb [8]. Instead of cutting the guard components from a plate, CFRP flat bars were employed. In this new guard design, CFRP flat bars with a width of 3 mm and a thickness of 1 mm were used. A sphere shape of guard can be made because the thin and flat CFRP bars can be bent easily when force is applied. As a result, the assembled guard protects all parts of the drone, as shown in Fig. 8.

Fig. 7. Example of pendant light guard frame [8]

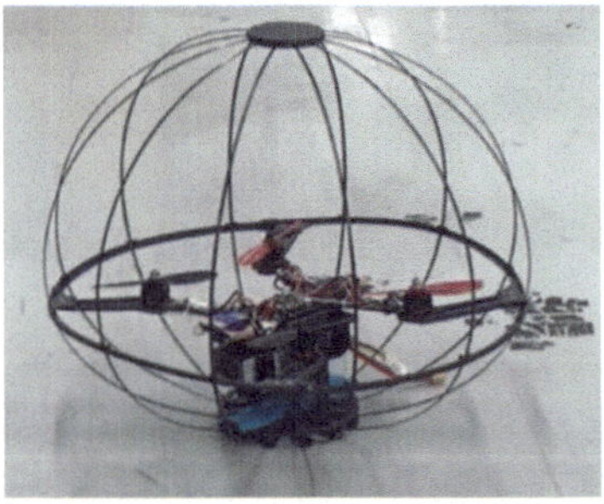

Fig. 8. Spherical guard made of CFRP

4.3 Guard Components

The guard components can be divided into four parts: upper plate, lower plate, middle frame, and CFRP flat bars, as shown in Fig. 9. Except for the CFRP flat bars, most parts were designed using the 3D CAD software, SOLIDWORKS, and the components were fabricated with a 3D printer. The filament of PA12-CF, a composite carbon fiber material known for its high strength and toughness, was used. Holes were made in the upper plate, the lower plate, and the middle frame components to fix the CFRP flat bars. The bars are inserted through the middle frame to fix their position and then assembled by inserting them into the holes in the upper and lower plates.

The guard should be removable, when adjusting or modifying the drone. Therefore, chemical adhesives were not used. In addition, it is impossible to make holes in the thin and flat CFRP bars. Consequently, rubber bands were used to secure the bars in the holes of parts such as the upper and the lower plates, as shown in Fig. 10. Furthermore, the guard was designed so that the center of gravity of the entire drone is positioned lower, with the upper part of the guard having a radius of about 230 mm and the lower part a radius of about 115 mm, as shown in Fig. 11.

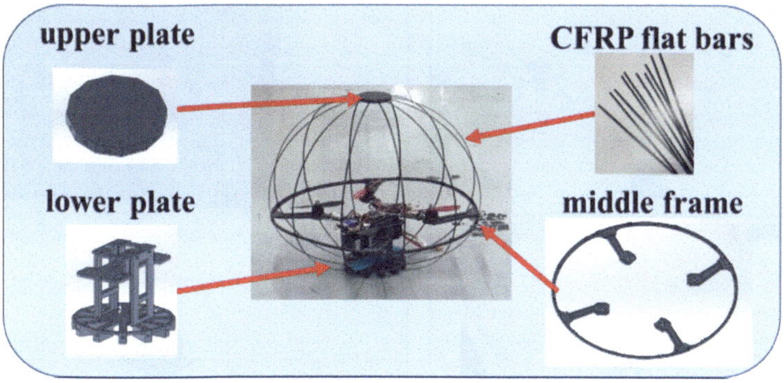

Fig. 9. Guard components

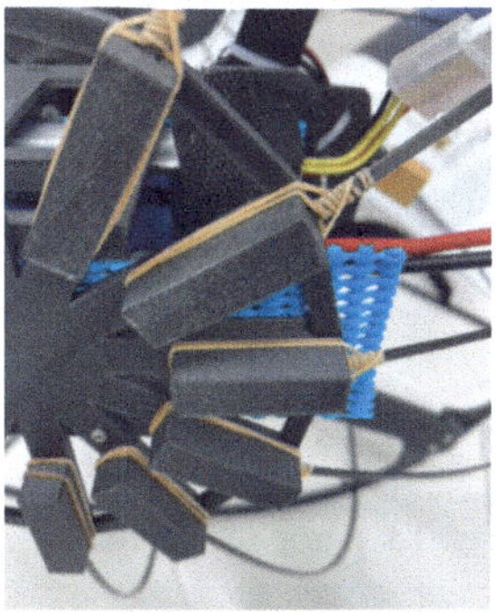

Fig. 10. Fixing CFRP rods to lower plate

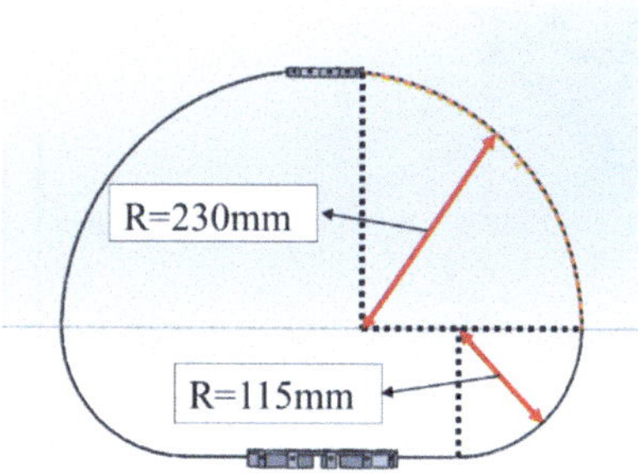

Fig. 11. Vertical section of the designed guard frame

5 Results of Experiments

5.1 Experiment to Test the Posture-Recovery Function

With the drone equipped with a CFRP guard, experiments were conducted to test the posture-recovery and continuous flight functions. First, an experiment was performed to assess the posture-recovery function. As shown in Fig. 12, the drone with the guard was initially placed upside down to verify whether it would automatically return to its original upright position. It was confirmed that the drone and battery act as weights and the spherical shape of the guard helps its easy rolling motion, then causing the drone automatically returned to its original upright posture. Next, as shown in Fig. 13, the drone with the guard was released from a high position and allowed to free-fall. In this case as well, it was confirmed that the drone automatically returned to its original upright posture as a result.

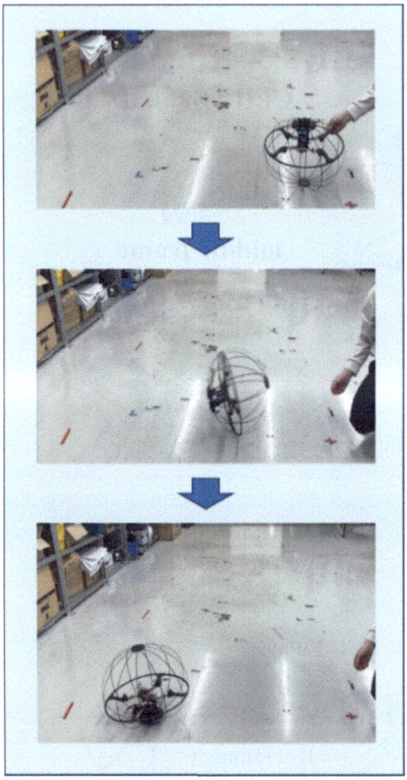

Fig. 12. Rolling the drone

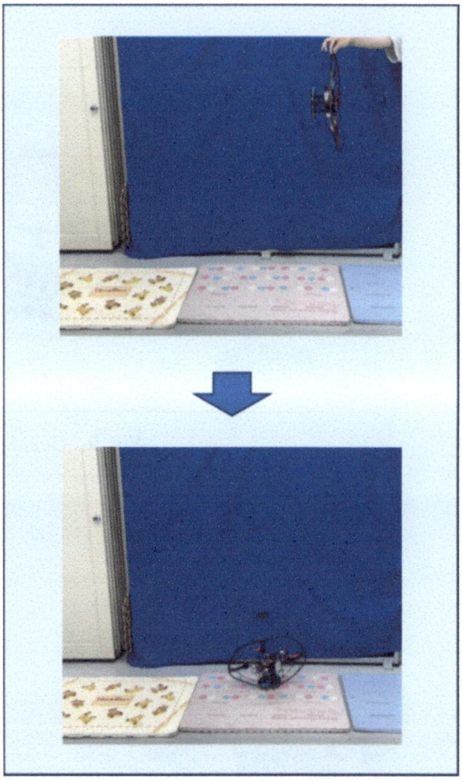

Fig. 13. Drop the drone

5.2 Experiment to Test Continuous Flight Function

Another experiment was conducted to verify whether the drone could continue flying even upon contact with a wall. As shown in Fig. 14, the drone equipped with the guard flew within a confined space. The drone was manually operated and made to collide with the walls from all directions; top, bottom, left, and right. The results confirmed that the drone was protected securely, and was able to continue flying without crashing, regardless of the direction of impact.

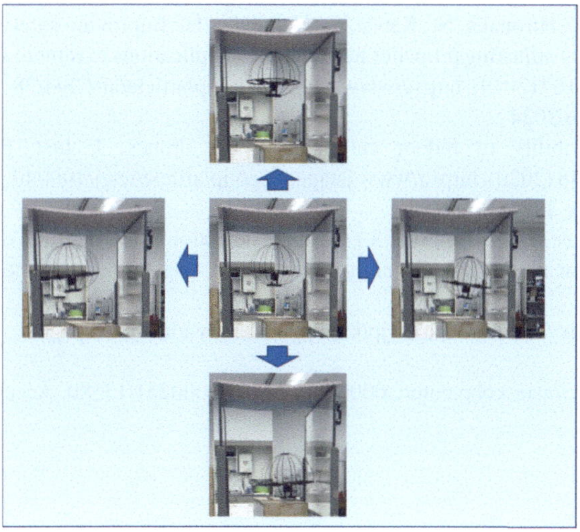

Fig. 14. Flight in confined spaces

6 Conclusion

In this study, a guarded drone that has continuous flight and posture-recovery functions were developed. By designing the guard in a spherical shape and lowering the vehicle's center of gravity, both functions were realized, inspired by the concept of the Okiagari Koboshi, a self-righting doll. Through experiments using the developed drone system, it has been confirmed that the drone can fly robustly even in collisions with walls and naturally return to its take-off posture and take off again, even after falling.

However, several challenges remain in this research. The first challenge is durability. Although the strength of the guard was improved by using CFRP material, it is still insufficient to withstand strong impacts, such as high-speed collisions or falls from significant heights. The second one is the size limitation of the drone. A small drone was used in this study, however a larger drone with other devices is needed for various applications.

References

1. Tatematsu, Y., Kawamura, I.: A consideration on support tool for pesticide application by drones. In: Proceedings for the Type II Workshop of the Japanese Society for Artificial Intelligence (2022). https://www.jstage.jst.go.jp/article/jsaisigtwo/2022/CCI-009/2022_07/_pdf/-char/ja. Accessed 14 Aug 2024
2. Ohno, K., Okada, Y., Harasima, S., Yokoe, M.: Unmanned aerial vehicle for supporting close visual inspection of bridges. J. Jpn. Soc. Precis. Eng. **83**(12), 1066–1070 (2017). https://www.jstage.jst.go.jp/article/jjspe/83/12/83_1066/_pdf/-char/ja. Accessed 14 Aug 2024
3. Naka, Y., Nakamura, Y., Hirasawa, G.: Cooperation with route movement learning of an AGV and obstacle detection of a drone. AIJ J. Technol. Des. **30**(74), 502–505 (2024). https://www.jstage.jst.go.jp/article/aijt/30/74/30_502/_pdf/-char/ja. Accessed 14 Aug 2024

4. Zhang, H., Li, H., Hironaka, N., Kawai, S., Nobuhara, H.: Improving safety and reducing size in multicopters by adjusting propeller height and its applications to remote sensing. Agric. Inf. Res. **28**(4), 157–167 (2019). https://www.jstage.jst.go.jp/article/air/28/4/28_157/_pdf/-char/ja. Accessed 14 Aug 2024
5. Sakai, Y.: Possibility of indoor automatic flying drones. J. Inst. Electr. Installation Eng. Jpn. 246–248 (2020). https://www.jstage.jst.go.jp/article/ieiej/40/4/40_246/_pdf/-char/ja. Accessed 14 Aug 2024
6. Hirumoto, M., Lee, J.H., Okamoto, S.: Obstacle detection using multiple distance sensors on a drone. Bachelor thesis, Department of Mechanical Engineering, Ehime University, Japan (2022)
7. HondaKids. https://www.honda.co.jp/kids/jiyuu-kenkyu/lower/21/page2/. Accessed 21 Aug 2024
8. Asia-kobo. https://asia-kobo.com/c/0000000124/0000000231/13580. Accessed 14 Aug 2024

Autonomous Flight and Path Planning for an Indoor Inspection Drone Using LiDAR

Yudai Nagata, Jae Hoon Lee(✉), and Shingo Okamoto

Graduate School of Science and Engineering, Ehime University, 3 Bunkyocho, Matsuyama, Ehime 790-0826, Japan
lee.jaehoon.mc@ehime-u.ac.jp

Abstract. In recent years, drones have become prominent in inspection works, largely due to their high mobility. This study focuses on the use of drones for the inspection of infrastructure in indoor environments, such as inside factories or industrial buildings. However, it is difficult to perform inspection tasks manually, so developing stable and effective methods using autonomous flight is essential. Therefore, a navigation algorithm for a drone system equipped with LiDAR was investigated to enable autonomous inspection in indoor environments where GPS is unavailable. In general inspection tasks, capturing images of the surroundings is the primary objective, thus the drone should fly near objects or walls while maintaining a certain distance and perpendicular to them. As a solution, a path planning algorithm using point cloud data from LiDAR was developed to automatically generate flight paths in an indoor environment. It generates the flight path including waypoints with coordinates and yaw angles based on pre-acquired 3D point cloud data. Also, validation experiments were conducted. Demonstrating the system's usability through two main tests: The First test was autonomous flight using LiDAR, and the second was generation of inspection paths in the experimental environment, followed by autonomous flights along these generated paths for inspection tasks. Furthermore, a 3D model was created using RGB and depth images data captured during the autonomous flights.

Keywords: Drone · LiDAR · Autonomous flight · Path planning · Inspection work

1 Introduction

Drones are capable of excellent mobility in the air and can hover in a fixed position with a flight mechanism. Furthermore, they do not need large take-off and landing areas. As a result, drones are used in various situations, infrastructure inspections, aerial photography, light shows and so on [1–3]. This study focuses on the use of drones for the inspection of infrastructure in indoor environments, such as inside factories or industrial buildings. However, it is difficult to operate drones manually since there are complex structures in buildings, therefore, autonomous flight control is necessary. Moreover, it is impossible to use GPS in indoor environments, a position estimation algorithm must be

© The Author(s), under exclusive license to Springer Nature Singapore Pte Ltd. 2025
J. Mo and P. Chutima (Eds.): ICMSC 2024, LNME, pp. 141–149, 2025.
https://doi.org/10.1007/978-981-96-5380-5_14

embedded in the drone. Therefore, in this study, a system with SLAM technology using LiDAR, capturing 360-degree 3D point cloud data, was developed. In addition, since capturing images of surroundings is mainly a primary task in drone-based inspection work, it is required to fly near objects or walls with a certain distance and heading angle perpendicular to them. To meet these conditions, we also developed a path planning algorithm based on previous research [4] that generates the appropriate waypoints and yaw angles for inspection in an indoor environment.

2 Inspection Procedure

The following section explains the inspection procedure with the drone system developed in this study. The inspection procedure has four main steps as follows.

- Manual flight for collecting rough point cloud data: A manual flight is conducted to collect rough point cloud data of the inspection area. During manual flight, the drone is operated in loiter mode and position control is supported by position estimation using FAST-LIO2 [5]. Here, it is essential to collect a uniform 3D point cloud of the inspection environment using LiDAR embedded on the drone. The rough 3D map is given as a result in this step.
- Generation of the flight path: The developed path planning algorithm generates a flight path based on the rough 3D point cloud data, obtained from the manual flight. Here, the path is for inspecting indoor target space, forming a contour near the walls and with the drone oriented perpendicularly to the walls for capturing images in the optimal pose.
- Autonomous inspection flight: The drone autonomously flies according to the generated flight path where the drone's position is estimated on the rough 3D map with the localization of FAST-LIO-LOCALIZATION algorithm [6, 7]. During the flight, the depth camera mounted on the drone captures both RGB and depth images of the inspection targets.
- Creation of 3D Model: After the autonomous inspection flight, a 3D model of the inspection target is constructed from the captured image data using Open3D algorithm that is an open-source library for handling 3D data [8].

3 Development of Drone System for Inspection

3.1 Drone Systems

A quadcopter with four propellers was employed as the aircraft for the inspection flight. The total weight of the drone is 2.28 kg, width is 570 mm, and a height of 255 mm. A 4S 7000 mAh lithium polymer battery was employed. A Pix32 v6 flight controller and an Intel NUC13pro computer were used for both manual and autonomous flight control. A Livox Mid-360 LiDAR was utilized to collect point clouds around the drone. Additionally, an Intel RealSense D455 depth camera was used to capture images of the inspection target objects during autonomous flight. An overview of the drone is shown in Fig. 1 and the system configuration is displayed in Fig. 2.

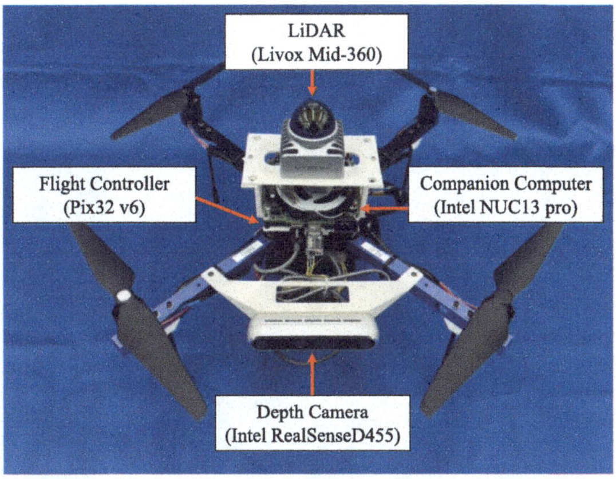

Fig. 1. Drone system developed in this research

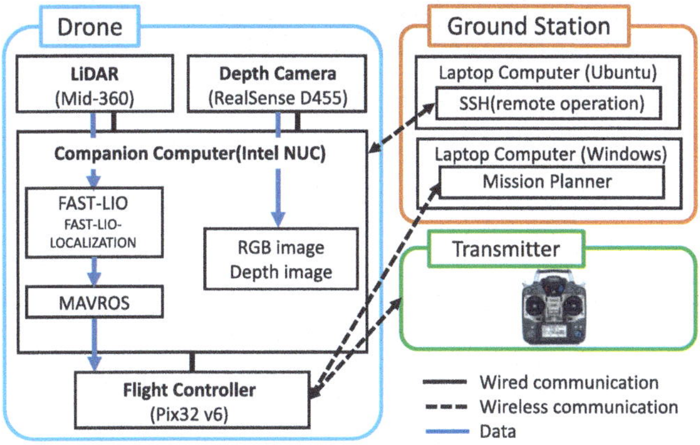

Fig. 2. System configuration of the developed drone

3.2 LiDAR

The Livox Mid-360 [9], employed as a LiDAR sensor to collect 3D point cloud data around the drone, is shown in Fig. 3. The combination of this LiDAR and SLAM technology enables 3D mapping and position estimation. In a previous study [7], position estimation was performed with a depth camera of the Intel RealSense D455[10] and a vision-based algorithm of ORB-SLAM2 [11]. However, when a camera is used for position estimation, the drone system fails to estimate the position information in cases of sudden rotation or fluctuation and becomes unstable due to the limited field of view of the camera. In this study, Mid-360, which has a wide sensing range of 360 degrees horizontal and 59 degrees vertical field of view and a maximum detectable range of 100 m, was used to improve the reliability of position estimation. The output frequency of the point cloud data is 10 Hz. The dimensions are 65 mm in length, 65 mm in width, and 60 mm in height, with a mass of 265g, compact and lightweight compared to general

LiDAR. It also incorporates a 3-axis accelerometer and a 3-axis gyroscope as part of its IMU module.

Fig. 3. Overview of LiDAR, Mid-360, Livox Co.

3.3 Depth Camera

The Intel RealSense D455 depth camera was used to capture both RGB and depth images of the inspection target during the inspection flight. The camera uses stereo vision to compute depth information, and in addition to acquiring depth data, it can also capture standard RGB images.

3.4 Companion Computer

The developed system uses an Intel NUC13pro minicomputer running the Ubuntu 20.04 LTS Linux operating system as a companion computer. It is responsible for activating LiDAR, cameras, executing SLAM and autonomous flight programs, and communicating with flight controller (FC) and ArduPilot [12] via the MAVLink protocol. Communication with the flight controller is carried out via serial communication. The algorithm of FAST-LIO2 was installed on the companion computer, and the acquired point cloud data from the LiDAR is used to estimate the drone's current position in GPS-denied environments. Additionally, FAST-LIO-LOCALIZATION, an algorithm for estimating the LiDAR's position within a 3D point cloud map, was implemented to determine the drone's current position. Furthermore, MavROS [13], an API used to communicate the estimated position information to the flight controller, was installed. The Robot Operating System (ROS) [14] framework was employed to link the algorithms and hardware such as LiDAR, cameras, and flight controllers for effective real-time control.

3.5 Ground Station

A laptop computer running the mission planner application of ArduPilot [15] was used as the first ground station. It communicates with the drone's flight controller, acquires information, and specifies the home position for autonomous flight. Another laptop running Ubuntu 20.04 was used as the second ground station. As this ground station cannot directly control the companion computer during flight, it connects via wi-fi to execute application programs in the companion computer remotely via SSH (Secure Shell).

4 Path Planning Algorithm

4.1 Algorithm Outline

The developed path planning algorithm generates a flight path based on the 3D point cloud data obtained from the manual flight. The flight path consists of a set of waypoints and yaw angles. The waypoints guide the drone to fly a certain distance from the wall and inspection target, while the yaw angles ensure the drone remains perpendicular to them for capturing images in optimal pose.

4.2 Preprocessing of Point Cloud Data and Creation of 2D Map

The 3D point cloud data includes unnecessary ceiling and floor information, which is removed to focus on detecting surrounding obstacles. Noisy points in the point cloud data are identified and removed. If a small number of points are found within a specified radius, they are considered noise and deleted, as deviating from the general point set of real objects. The preprocessed 3D point cloud data is then extracted within a specified height range and plotted on a 2D plane.

4.3 Extraction of Wall Shape

The contour of walls and obstacles in the target inspection space is identified using the plotted 2D map of each layer. Firstly, a Delaunay diagram is constructed to obtain adjacent point data from the 2D point cloud using Delaunay triangulation [16]. Delaunay triangulation a plane with multiple points in a way that maximizes the minimum angle of the triangles connecting each point. This ensures adjacent points are connected without overlap. In this study, the Delaunay triangulation program from the Scipy library was used. Subsequently, the alpha shape, a sub-complex of the Delaunay triangle, is created based on the Delaunay diagram [17]. The radius of the circumscribing circle of each triangle is compared to a pre-defined value. If the calculated radius is smaller than the set value, the edge is added to the list, contributing to the alpha shape. The radius of the circumscribing circle of a triangle formed by three points, r, is calculated using the following equation.

$$r = \mathrm{abc}/4\mathrm{A} \tag{1}$$

where, a, b, and c represent the lengths of the triangle's sides, and A represents the area. Any duplicate edges already present in the list are removed. This process extracts only the contour of the wall surface, as shown in Fig. 4. The green points represent the point cloud data, and the black points represent the alpha shape as the contour of the object surface.

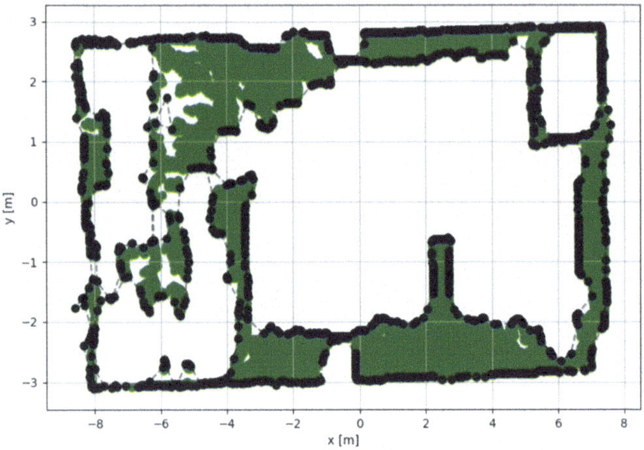

Fig. 4. An example of 2D point cloud map with alpha shape

4.4 Creating Offsets

The inspection flight path should be close to the wall, with a specific offset distance from the wall contour. Offset points for the flight path are generated as follows. First, start and goal points of the flight path are set at arbitrary locations on the 2D map. Then, a set of edges along the wall is extracted from a point near the start point to a point near the goal point. Each point along the edge set is shifted by a specified distance from the wall surface, creating offset points. Lines connecting the offset points intersect, the order of the points is adjusted to eliminate the intersection.

4.5 B-Spline Interpolation

A smooth trajectory is necessary for the inspection flight, so B-spline interpolation was applied to the offset path. A B-spline curve is a smooth curve defined by control points and a knot vector.

It has continuous characteristics and modifying some of the control points does not affect the entire curve. However, the generated curve does not precisely pass through the control points [18].

4.6 Calculation of Yaw Angle

To capture proper images during the inspection flight, the camera's heading direction (yaw angle) must be provided along with the waypoint coordinates. The yaw angle is calculated from a nomal vector of the B-spline curve. In this study, a path for inspection flight was generated in the right half of the environment, shown in Fig. 5. The direction of arrows at each point indicates the drone's yaw angle.

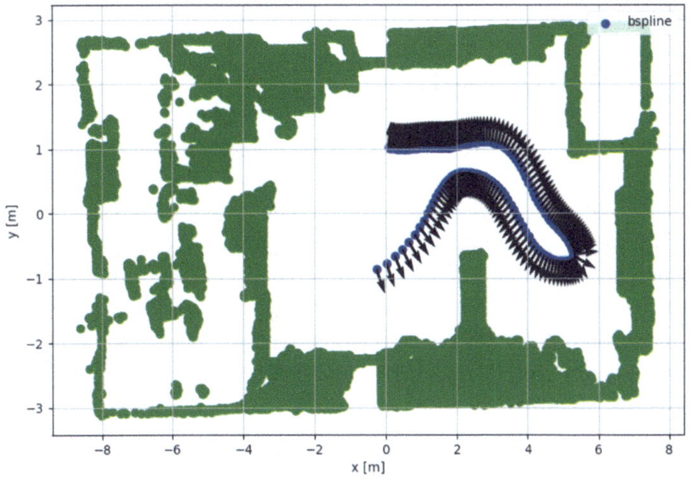

Fig. 5. Example of autonomous inspection path generated by the proposed algorithm for inspection right half space of the room

5 Demonstration Experiment

5.1 Autonomous Flight Using LiDAR

The developed drone was tested to confirm position estimation capability on the pre-constructed rough 3D map. Additionally, the flight control performance was tested through a series of experiments. Aplanned U-shaped path was created for autonomous flight experiment, and the drone flew along the path. The input path and the drone's flight trajectory are represented by green and red lines, respectively, in Fig. 6. It was observed

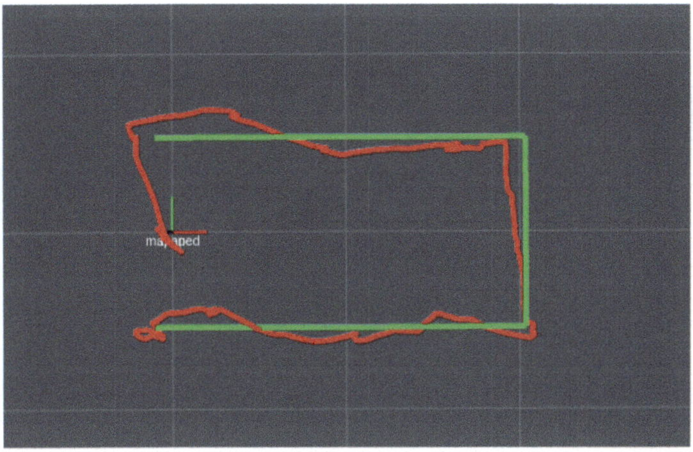

Fig. 6. Planned path (green line) and flight trajectory (red line) during autonomous flight experiment

that the drone followed the planed path well, although some position errors occurred dring the flight.

5.2 Autonomous Inspection Flight

Autonomous inspection flight experiments were carried out using paths generated by the path planning algorithm. First, a path at a height of 0.5 m was generated on the first layer, using point cloud data extracted between the heights of 0.3 m and 0.7 m. Next, a path at a height of 1.0 m was generated on the second layer, using point cloud data extracted between the heights of 0.8 m and 1.2 m. These two paths were integrated to create a two-stage flight path, and the drone flew according to this path during the inspection flight experiment. Figure 7 shows the generated path in green and the flight trajectory in red. Additionally, Fig. 8 shows a 3D model constructed from RGB and depth images captured during the inspection flight.

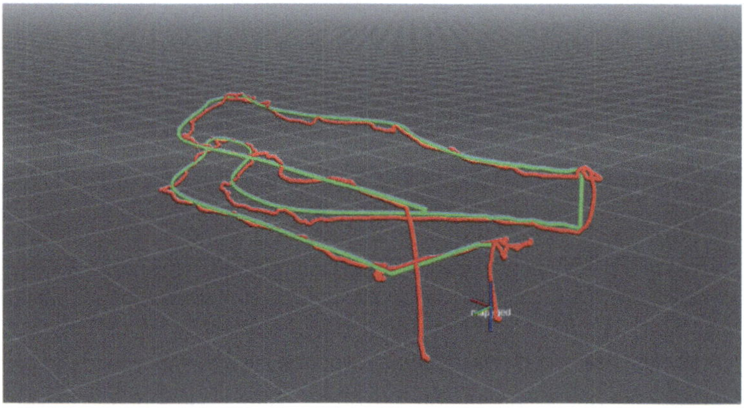

Fig. 7. Planned path (green line) and flight trajectory (red line) during autonomous inspection flight experiment

Fig. 8. 3D model constructed from the images captured during autonomous inspection flight experiment

6 Conclusions

In this study, an algorithm to navigate the drone for indoor inspection task was developed. By employing LiDAR and SLAM algorithms on the drone, robust localization to provide current position information was achieved. Additionally, path planning for autonomous inspection flights was realized based on a previously collected rough 3D map. The developed algorithm generated a flight path and yaw angle to capture images at the optimal angle. The proposed algorithm was tested through experiments with the developed drone in an indoor laboratory room, and its feasibility was demonstrated. In future work, the algorithm and the drone system will need to be modified for larger, real-world environments such as industrial buildings.

References

1. Seo, J., Duque, L., Wacker, J.: Drone-enabled bridge inspection methodology and application. Autom. Constr. **94**, 112–126 (2018)
2. Nar, D., Kotecha, R.: Optimal waypoint assignment for designing drone light show formations. Results Control Optim. **9**, 100174 (2022)
3. Vanegas, F., Gaston, K.J., Roberts, J., Gonzalez, F.: A framework for UAV navigation and exploration in GPS-denied environments. In: IEEE Aerospace Conference 2019, pp. 1–6. IEEE (2019)
4. Mizuta, Y., Lee, J.H., Okamoto, S.: Path planning with 3D point cloud for indoor inspection drone. In: The Twenty-Ninth International Symposium on Artificial Life and Robotics 2024 (AROB 29th 2024), pp. 521–525 (2024)
5. Xu, W., Cai, Y., He, D., Lin, J., Zhang, F.: Fast-LIO2: Fast direct LiDAR-inertial odometry. IEEE Trans. Robot. **38**(4), 2053–2073 (2022)
6. FAST_LIO_LOCALIZATION Homepage. https://github.com/HViktorTsoi/FAST_LIO_LOCALIZATION. Accessed 15 Aug 2024
7. Shan, T., Englot, B., Meyers, D., Wang, W., Ratti, C., Rus, D.: LIO-SAM: tightly coupled LiDAR inertial odometry via smoothing and mapping. In: IEEE/RSJ International Conference on Intelligent Robots and Systems (IROS) 2020, pp. 5135–5142. IEEE (2020)
8. Zhou, Q. Y., Park, J., Koltun, V.: Open3D: a modern library for 3D data processing. arXiv preprint arXiv:1801.09847 (2018)
9. Livox Mid-360 Homepage. https://www.livoxtech.com/jp/mid-360. Accessed 15 Aug 2024
10. Intel RealSense D455 Homepage. https://www.intelrealsense.com/depth-camera-d455. Accessed 15 Aug 2024
11. Mur-Artal, R., Tardós, J.D.: ORB-SLAM2: an open-source SLAM system for monocular, stereo, and RGB-D cameras. IEEE Trans. Robot. **33**(5), 1255–1262 (2017)
12. ArduPilot Homepage. https://ardupilot.org. Accessed 15 Aug 2024
13. MavROS Homepage. https://wiki.ros.org/mavros. Accessed 15 Aug 2024
14. ROS Homepage. https://wiki.ros.org/ja. Accessed 15 Aug 2024
15. Mission planner Homepage. https://ardupilot.org/planner. Accessed 15 Aug 2024
16. Shewchuk, J. R.: Delaunay refinement mesh generation. Carnegie Mellon University (1997)
17. Alpha shapes Homepage. https://graphics.stanford.edu/courses/cs268-11-spring/handouts/AlphaShapes/as_fisher.pdf. Accessed 15 Aug 2024
18. Bspline Homepage. https://www.cl.cam.ac.uk/teaching/1999/AGraphHCI/SMAG/node4.html. Accessed 15 Aug 2024

Author Index

© The Editor(s) (if applicable) and The Author(s), under exclusive license
to Springer Nature Singapore Pte Ltd. 2025

J. Mo and P. Chutima (Eds.): ICMSC 2024, LNME, pp. 151–152, 2025.
https://doi.org/10.1007/978-981-96-5380-5

The manufacturer's authorised representative in the EU is Springer
Nature Customer Service Centre GmbH, Europaplatz 3, 69115 Heidelberg,
Germany. If you have any concerns regarding our products, please
contact ProductSafety@springernature.com

Printed and bound by CPI Group (UK) Ltd, Croydon, CR0 4YY
01/10/2026
02225944-0013